Encounters in Virology

Teri Shors
University of Wisconsin–Oshkosh

D0878614

JONES & BARTLETT
LEARNING

World Headquarters
Jones & Bartlett Learning
5 Wall Street
Burlington, MA 01803
978-443-5000
info@jblearning.com
www.jblearning.com

Jones & Bartlett Learning books and products are available through most bookstores and online booksellers. To contact Jones & Bartlett Learning directly, call 800-832-0034, fax 978-443-8000, or visit our website, www.jblearning.com.

Substantial discounts on bulk quantities of Jones & Bartlett Learning publications are available to corporations, professional associations, and other qualified organizations. For details and specific discount information, contact the special sales department at Jones & Bartlett Learning via the above contact information or send an email to specialsales@jblearning.com.

Production Credits
Chief Executive Officer: Ty Field
President: James Homer
SVP, Editor-in-Chief: Michael Johnson
SVP, Chief Marketing Officer: Alison M.
  Pendergast
Publisher: Cathleen Sether
Senior Acquisitions Editor: Erin O'Connor
Editorial Assistant: Rachel Isaacs
Production Manager: Louis C. Bruno, Jr.
Senior Marketing Manager: Andrea DeFronzo
V.P., Manufacturing and Inventory Control:
  Therese Connell

Composition: M&M Composition, LLC
Cover Design: Kristin E. Parker
Rights & Photo Research Associate: Lauren
  Miller
Cover Image: Organism © BioMedical/
  ShutterStock, Inc.; puzzle photo © Stillfix/
  ShutterStock, Inc.
Printing and Binding: Edwards Brothers Malloy
Cover Printing: Edwards Brothers Malloy

Library of Congress Cataloging-in-Publication Data unavailable at time of printing.
ISBN: 978-0-7637-7349-6

6048

Printed in the United States of America
16 15 14 13 12   10 9 8 7 6 5 4 3 2 1

This book is dedicated to the late
Elaine (Motschke) Gross, my mother.

Contents

Preface

Many individuals find viruses fascinating. Viruses have starred in Hollywood movies such as *Contagion*, *Rise of the Planet of the Apes*, *Zombieland*, *Resident Evil*, *I Am Legend*, *28 Days Later*, *28 Weeks Later*, *The Stand*, *Outbreak*, *The Andromeda Strain*, *Quarantine*, *Twelve Monkeys*, *Mulberry Street*, *Flu Birds*, *Pathogen*, *Carriers*, *The Omega Man*, *The Last Man on Earth*, *The Cassandra Crossing*, and *Cabin Fever*. They have also starred in the recent popular AMC TV series *The Walking Dead* and the BBC series *Survivors*. The viruses in these horror or drama/thriller movies are scarier than the H1N1 virus of the 2009 influenza pandemic. How is it that these microscopic infectious agents can cause disease or impact ecosystems? Their wily and versatile nature is intriguing. New and emerging viral diseases continue to challenge humans and set the stage for future pandemics.

Encounters in Virology complements the current "whatdunit" two-volume set of books titled *Encounters in Microbiology*. The majority of accounts in *Encounters in Microbiology* are about infectious diseases caused by bacterial pathogens. *Encounters in Virology* focuses on viral or prion diseases. The majority of cases are inspired by true events documented in ProMED or primary literature. For most of these "encounters," the characters, certain events, and dialogues are fictional. Unless otherwise noted, the names are fictional.

"The Steps Used When Diagnosing and Treating a Patient" found in the original *Encounters in Microbiology* volumes is included to familiarize the reader with the tools used to diagnose and treat the patient. References and a set of questions accompanies each viral encounter and patient diagnosis carried out by clinicians. Many of the questions stress critical thinking skills. A glossary and index are also included. These cases can complement the textbook *Understanding Viruses, Second Edition*, and other general virology and microbiology textbooks.

Acknowledgments

I thank Developmental Editor Molly Steinbach, Acquisitions Editor Erin O'Connor, Editorial Assistants Rachel Isaacs and Agnes Burt, photo researcher Lauren Miller, and the production team at Jones & Bartlett Learning, especially Production Manager Louis Bruno for making this possible. I also thank Rhonda Mesko and Jennifer Richer for critically reviewing this manuscript, Brian Ledwell for illustrating the figures, and Jeffrey Pommerville for supporting the need for more case studies based on viral applications.

I am most grateful to those individuals on the sidelines of the manuscript. John Cronn's teaching, advice, and kindness cannot go without mention. Getting to know Robert I. Krasner during this time and his humorous wit is something I will always treasure. Darin Reiger's companionship, love of zombies, vampires, SpongeBob Squarepants, the Sneeze game, and the *Walking Dead* series created the perfect writing environment. Becca "Doodle" Reiger's laughter and intellectual playfulness was treasured each and every day. Lastly, I owe my deepest gratitude to my mother, Elaine (Motschke) Gross, and dedicate this book to her. She was and still is my fuel.

Introduction

Virology is an exciting, fast-paced field of biology. It is a field that offers the opportunity to work in a variety of settings. It can be a demanding science, requiring virologists to possess a well-rounded background in mathematics, physics, general microbiology, chemistry, molecular and cellular biology, and ecology to fully appreciate the structure of viruses, reproduction, and their interactions with their hosts and environment. Viruses hold the key to understanding biological systems.

Viruses infect essentially all life forms. Historical writings about viruses can be traced back several thousand years. Many viruses mutate rapidly, allowing them to adapt to changing environments. All that is needed is contact between an infected individual and a susceptible person to spread a fire-like virus in a population. This vulnerability is what fascinates and lures someone into using a killer virus as a subject in a Hollywood screenplay or best-selling novel, becoming a virologist, or simply reading these practical applications.

The following vignettes are fictional stories based on actual viral cases or outbreaks. I hope that these contemporary,
real-world virology encounters will increase your interest
and desire to learn more about viruses. Some of you may be
inspired to become virologists, epidemiologists, physicians,
or medical, science, or other types of writers or journalists.

Enjoy being a virus detective!

Teri Shors, PhD

The Steps Used When Diagnosing and Treating a Patient

The identification of the nature and cause of a patient's illness or disorder by a clinician is called a diagnosis. When the ill individual comes to the clinician's office or medical clinic, a series of diagnostic steps (**Figure A**) are set in motion. This includes a patient interview and an evaluation of the patient's reported symptoms, the physical examination findings determined by the clinician, the results of various laboratory and medical tests, and any other procedures pertinent to the investigation. If the patient's illness appears to be caused by an infectious agent, then the investigation may need to identify the causative agent and characterize the severity of the infection. When all this information has been evaluated and the clinician has reached a diagnosis, she or he can offer a prognosis, a prediction of the likely outcome of the disease. From this, a treatment procedure can be started and preventative (and possible public health) action initiated.

Because we are specifically interested in infectious diseases and disorders, the first step is to determine the **exposure history** of the patient.

1. Exposure History

The clinician will conduct an exposure history interview as part of the patient's overall personal and family history. Presentation to a clinician includes the patient's current illness and an oral report of his or her subjective symptoms. With regard to exposure, the clinician must cover the following topics when interviewing and examining the patient: can the patient determine the time of onset; can the patient pinpoint the time and place of exposure (e.g., home, work, recent domestic/international travel); has the patient had past infectious diseases, vaccinations, or immunological impairments; can the patient identify possible exposure sources (other humans, animals, foods, or environment).

The clinician's interview may indicate obvious signs or symptoms. For example, a child with a case of chickenpox (small, teardrop-shaped, fluid-filled vesicles on the torso) would be quite obvious to an examining pediatrician. In this case, little, if any, further investigation would be needed for a correct diagnosis, and the prevention of spread can be discussed. If this represents the beginning of an outbreak or

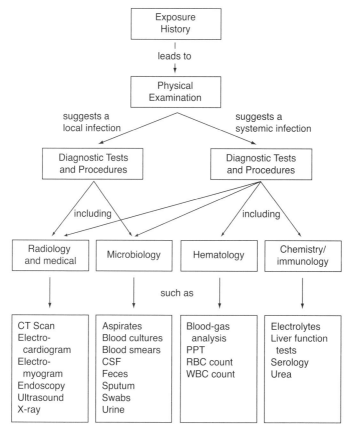

Figure A. Flow diagram of the diagnosis of an infection. CT = computed tomography; CSF = cerebrospinal fluid; WBC = white blood cell; RBC = red blood cell; PTT = partial thromboplastin time. (Modified from Greenwood, D., Slackl, R. C. B., and Peutherer, J. F. *Medical Microbiology*, 16th edition. London: Churchill-Livingston, 2002.)

epidemic, however, then more specific public health measures may need to be initiated and a report made to state and national medical authorities. In other cases, signs and symptoms may not be so direct. The presence of a headache, fever, and malaise for two days could be symptoms for chickenpox as well as for a large variety of viral and bacterial infections.

In the interview with the patient, a review of body systems may uncover other parts of the body being affected by the disease. For example, coughing and shortness of breath may indicate respiratory system involvement. On the other hand, a burning on urination would suggest a urinary tract infection. All are part of the disease detective work done to locate the physical site of symptoms or what is called an anatomical diagnosis. Such a diagnosis may allow the clinician to narrow the list of possible infections or infectious agents.

2. The Physical Examination

As part of the physical examination, the clinician does a systematic examination of the patient, taking a blood pressure reading, measuring heartbeat, and measuring body temperature. Specific emphasis is placed on the part of the body affected by the illness. For example, the throat, chest, and lungs are examined if a respiratory system infection is suspected.

The examination may allow the clinician to narrow the possibilities of diseases and/or infectious agents that would fit the clinical findings. The exposure history and physical exam may lead to the determination if the infection is local, such as the lungs or urinary tract, or systemic, involving several tissues/organs in the body. The clinician may then be ready to make a differential diagnosis, which narrows down the potential diseases to just those few that fit the clinical findings. If the presenting symptoms in a 45-year-old patient are a three-week cough, fever, chest pain, and coughing up blood or sputum, a differential diagnosis may include several respiratory infections but primarily tuberculosis (TB) as the cause. On the other hand, if the patient remembers a tick bite and has an expanding red rash at the bite site, the differential diagnosis is almost certainly Lyme disease, as few other arthropod-borne diseases have these specific signs.

3. Diagnostic Tests and Procedures

As illustrated by many of the stories in this book, the first two stages—exposure history and physical examination—are carried out rather quickly, often on the initial interview. However, the clinician may order one or more specific diagnostic tests to narrow down the short list of possible infections. Such tests or procedures can take some time and may be expensive. This is one reason why a clinician might attempt a final diagnosis through the physical examination or by using a minimal number of "standard" tests. For example, if TB is suspected, a chest x-ray or tuberculin skin test may be ordered to confirm the diagnosis. In addition, some diagnostic procedures may be noninvasive while others are invasive. Thus, comfort to the patient must be considered when diagnostic tests or procedures are being considered. Also, if someone such as a general practitioner is treating a patient, she or he may consult with an infectious disease specialist to obtain an expert opinion. Even Google searches have been used to help with patient diagnoses!

In some cases it may not be necessary to identify the specific pathogen as part of the diagnosis. For example, if all the diseases or agents identified by a differential diagnosis would be treated in the same way, or not at all (e.g., common cold, measles), there probably is no need to identify the pathogen. On the other hand, sometimes it is necessary to identify the actual causative agent of the infection. This etiological diagnosis may be important especially if it is a particularly dangerous disease. For example, if the patient is infected with HIV or hepatitis C virus, it is necessary to treat the patient with antivirals and to monitor their viral loads. If the viral load increases, a different regime of antiviral treatment would be needed to treat the patient.

During the differential and etiological diagnoses, the clinician needs to be aware of a number of important epidemiological issues. The clinician needs to know if other individuals are at risk. Do particular behaviors (traveling, working in crowded places, etc.) expose one to the disease, and has the patient engaged in these behaviors? What is the geographical distribution of the disease, and has the patient been in these locales? Have there recently been additional cases reported locally? Has the patient been immunized, if possible, against this disease? During the patient interview and examination, many of these questions may be answered by the patient, assuming (which one often cannot) that the patient's ability to "self-report" is honest and accurate.

Although diagnoses and diagnostic tests obviously demand good judgment on the part of the clinician, for some conditions, written flow diagrams called decision trees (or algorithms) exist for making diagnostic decisions and for treating the patient; in other words, "If the patient has this, do the following test." Medical and health insurance companies often use diagnosis and treatment algorithms. An example of a decision tree for evaluating a patient suffering from diarrhea of an unknown etiology is illustrated in **Figure B**. As you can see, they can be quite extensive.

4. Treatment

Once the clinician has reviewed all the clinical information and diagnostic tests, hopefully a correct diagnosis can be made and the prognosis issued. Then treatment can begin. *Note:* Often as a precautionary measure, treatment may begin while diagnostic tests are being run.

There are two possible treatment scenarios. In **symptomatic treatment**, the clinician treats symptoms, such as pain, fever, cough, or muscle aches accompanying the underlying disease. Pain relievers, antihistamines, or cough suppressants may be prescribed for something like a cold or the flu. These treatments are simply supportive, making the patient feel better without influencing the final outcome or progression of the disease.

In a **specific treatment**, the clinician is specifically treating the diagnosed disease and hopefully affecting the final outcome. Typical specific treatments might be prescribing an antibiotic for a sinus infection or several antibiotics for something like TB. Antiviral agents, such as acyclovir, might be prescribed for shingles, although in actuality that is only treating the symptoms. Hopefully through proper treatment, the patient will progress through a period of disease decline and complete convalescence. However, less optimistic outcomes due to deadly pathogens sometimes occur, as some of the stories will describe.

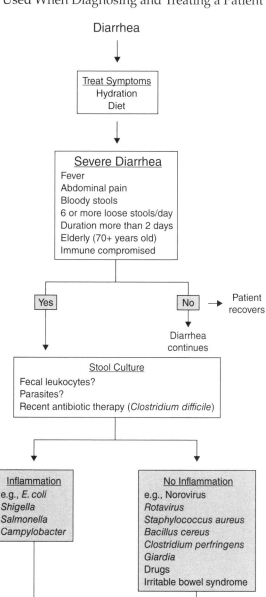

Figure B. Assessment and evaluation algorithm used for patients suffering from diarrhea of unknown etiology.

Chapter 1

Warts Running Wild

This encounter illustrates a very rare condition that had been left untreated for nearly 20 years. A confirming diagnosis was needed to design a treatment to improve the patient's quality of life. Improved technology often plays an important role in identifying what infectious disease agent(s) might be present in biopsy tissue. In this case, tests used to determine the immune status of the patient also helped explain why his condition was so severe.

American dermatology expert Dr. Gaspari was tired but eager to arrive in Jakarta, Indonesia. He was on his way to investigate the case of a 35-year-old Indonesian man with a very rare skin condition, about whom he had learned by watching a Discovery Channel documentary. The people in the remote village where the man, Dede, lived referred to him as the "Tree Man of Java." From the program, Dr. Gaspari observed that Dede had cauliflower-like tumors all over his body. His hands and feet looked like branches of trees growing from the surface of his skin. Dr. Gaspari was anxious to biopsy these growths, hoping that this would help him determine the cause of growths, as well as a treatment that would cure them, or at least improve Dede's quality of life.

He contacted the Discovery Channel and was provided with contact information for the Indonesian journalist who first covered Dede's story. After months of communication and planning, Dr. Gaspari was finally on his way to meet the "Tree Man" (**Figure 1**).

The journalist who had written the first story about Dede and his condition picked Dr. Gaspari up at the Jakarta airport. They traveled by car, by boat, and on foot to Dede's home, located in a remote jungle village.

Figure 1. Dede, the "Tree Man." (Drawn by Brian Ledwell.)

When the two finally met, Dr. Gaspari smiled and said, "Hello Dede, it is great to finally meet you. I would like to help you as much as I can. I believe there must be a medical treatment to help you."

Dede seemed nervous but excited to meet Dr. Gaspari. He explained that his biggest fear was that his two children would come down with his disease.

Dr. Gaspari needed to do a full body examination of Dede, but first he needed to learn more about Dede's history. This was an unusual case. He had never seen anything like it, nor had it ever been described in any medical textbook.

Dede began to speak shyly. "When I was 15 years old, I cut my knee. After that, the first **warts** started to grow. I used to work in construction, and I was a fisherman. I was married and had two children. But as time went by, the warts got worse and worse and I had to stop working when I was in my 20s. Then my wife abandoned me. I worry that the warts will cover my whole body. I worry that this disease will affect my children."

Dr. Gaspari observed the growths around his eyes, nose, and mouth.

Dede sighed, "I cannot care for my children financially. I cannot even touch my children. My sister is looking after them."

Dede looked serious and frightened. Dr. Gaspari sensed he was depressed and felt isolated by his condition. He may also have been malnourished. What he had was a unique, seemingly unstoppable infection. He listened carefully to what Dede had to say, and then gently asked, "How do your hands and feet feel?"

"Heavy," Dede replied. "I cannot use my hands for anything. I have not been able to write for over 10 years. I cannot feed myself. I have to drink through a straw. I cannot bathe myself. And it is very painful to walk." His eyes saddened as he spoke. "My brother-in-law helps care for me each day. He bathes and washes me. I need a special tailor to make clothes that will fit over my horned hands and feet."

Many thoughts ran through Dr. Gaspari's mind. What could be the cause of Dede's malady? Was his immune system compromised in any way? How is this man able to survive in this community? From Dede came an unsolicited soft-spoken reply, "I work in a circus troop, but this income is not enough to support my children."

Dr. Gaspari asked, "Have you had any treatment for this condition?"

"Doctors tried to remove some of the warts several years ago, but they grew back even faster."

The journalist chimed in, "The warts grow about 5 mm a month."

Dr. Gaspari knew this was a very challenging case. Dede had a life-threatening condition that required medical attention, but until a diagnosis could be made, proper treatment could not be possible. The wart infection had been allowed to flourish for 20 years without treatment. Dr. Gaspari also needed to figure out how well Dede's immune system was working in order to make a diagnosis. If his immune system was compromised, could it be boosted with drug therapy? Would that help bring the warts under control?

He began a full body examination, closely observing the warts. He was concerned that Dede might have skin cancer or be at risk for developing it soon. There were also infectious agents, like fungi and viruses, which could cause the formation of warts. He needed to take **biopsies** of the warts as a first step to figuring out what was causing them to form. "Dede, I need to remove some warts so that I can study them in my research laboratory in the United States." said Dr. Gaspari.

Dede said, "I want you to treat me."

Dede's home was small and cramped, and the conditions were hot and humid. There were many flying insects in and outside of Dede's residence. Flies were landing on Dede's warts and he could not swat them away.

Before the biopsy procedure began, Dede lay in the middle of the floor on a blanket. His relatives surrounded him. They quietly and intently watched the interactions between Dede and Dr. Gaspari. None of them could speak English. The journalist acted as an interpreter for them all.

Dr. Gaspari opened his medical kit and donned a pair of latex gloves. He opened a package containing a sterile needle and a vial of anesthetic. "You will feel a sting," Dr. Gaspari said. He injected an anesthetic under the skin, numbing the area around a large wart. Using a scalpel, he surgically removed a wart and put it into a specimen vial. The area around the removed wart began to bleed more than normal. It took more than 30 minutes to stop the bleeding. This was a cause for concern. Dr. Gaspari decided to remove smaller warts in the hopes that they would result in less bleeding. By the time he was finished, he had surgically removed a total of four warts.

Each time a wart was removed, a strong odor emanated. Dr. Gaspari considered that there was likely some bacterial colonization around the warts because it is very hard to keep them clean and free of microbes. As Dr. Gaspari left for the evening, he found himself concerned about the excessive bleeding caused by removing the warts. He worried that this would be a source for additional bacterial infection and didn't know if Dede's body could fight off such an infection. He was very concerned that he may have a weak immune system.

The next day, Dr. Gaspari returned. It was raining, hot and humid. The flying bugs were irritating.

"How are you doing, Dede?" said Dr. Gaspari with a smile. Dede smiled shyly. Dr. Gaspari analyzed the biopsy areas on Dede. "Your biopsy sites look good. You are healing. Dede, I need to figure out if your immune system is working right. In order to do that, I am going to inject some yeast cells under your skin. If the area turns red and swollen, it means your immune system is responding to foreign cells—that your immune system is normal."

Twelve hours after Dr. Gaspari injected the yeast, Dede's reaction was very weak. This was not a surprise to Dr. Gaspari. It affirmed his suspicions that Dede's immune system was impaired.

"Typical warts do not become this consuming. Warts come and go in individuals with healthy immune systems," explained Dr. Gaspari. "Because your body cannot fight off what is causing the warts, medical treatment is necessary to reduce them. I will need a blood sample from you and your family to help figure out what is causing these warts so that we can treat them with a drug. Blood samples will determine if your relatives are hidden genetic carriers of a disease or have a tendency towards a disease, which may have been passed to you, which is causing these growths. It will also help me learn more about your immune system."

"Do what you need to do," replied Dede.

Subsequently, Dr. Gaspari drew blood samples from Dede, his parents, sister, and children. As he took the samples, he continued to explain. "Surgically removing all of the warts will not solve the problem. A combination of surgery and chemotherapy may provide the best treatment so that Dede can live a more normal life."

"I want to be able to return to fishing and take care of my children," said Dede.

Taking the samples he had collected, Dr. Gaspari returned to his university laboratory in Baltimore, Maryland. He spent 2 months studying the biopsy samples, looking for agents like fungi, poxvirus, herpesvirus, or papillomavirus that could be the source of infection. Thorough tests were done on the blood samples. He was surprised to discover Dede had a nearly nonexistent immune system and a very low white blood cell count. He wondered if Dede could have **AIDS**.

He was also able to detect a papillomavirus in the warts. While papillomaviruses are very common throughout the world, the severity and extent of Dede's infection was unique. His first assessment was that Dede needed to be treated with a retinoid such as acetretin, which is a laboratory-derived form of vitamin A to reduce the warts. Retinoids work by regulating the growth of epithelial skin cells, which had raged out of control in Dede's condition. Another possibility was using interferon to help boost Dede's immune system to fight the viral infection.

Dede received worldwide attention. Once a diagnosis was made, local doctors began surgically removing Dede's warts at a hospital in Bandung. They removed over four pounds of warts. Unfortunately, there were complications. The bones of his fingers and toes were weak from so many years of disuse. He was treated with anti-osteoporosis drugs and calcium to strengthen his bones. He underwent skin grafts to heal the areas most affected by the warts.

After several months, Indonesian doctors began working with Dr. Gaspari to treat Dede chemotherapeutically. Dede received vitamin A and an antiviral drug that was only available in the United States. Today, Dede is improving. He has hope.

Update

Dede was in the hospital over 9 months recovering from numerous operations. Although unconfirmed, there were rumors that Dede also suffers from tuberculosis and hepatitis B. Today he can use his hands and walk. He will likely need more surgical procedures because the growths will continue to reappear in other locations of his body or regrow.

Questions to Consider

1. How could a laboratory technician discern that the warts were caused by a human papillomavirus and not a human herpes 8 virus or a monkeypox infection?

2. The disease described was left untreated for nearly 20 years. Could early treatment have prevented this condition? Why or why not?

3. How many types of papillomaviruses are there?

4. There are three distinct categories of papillomaviruses: low-, intermediate-, and high-risk types. Explain what this means.

5. Why is the disease in this encounter rare in humans?

6. If Dede also suffers from tuberculosis and/or hepatitis, how might this complicate his treatment?

7. Would the **GARDASIL**® or **CERVARIX**® **vaccines** have prevented this infection in Dede? Why or why not?

8. Research acetretin. For what types of maladies is acetretin used as a treatment option? Are there side effects to this drug?

9. Research interferon. When is interferon used as a treatment option? Does it have side effects?

10. Why was it difficult to remove Dede's warts surgically? Why was a combination of treatments used in Dede's case?

References

Alisjahbana, B., et al. 2010. Disfiguring Generalized Verrucosis in an Indonesian Man with Idiopathic CD4 Lymphopenia. *Arch Dermatol* 146:69–73.

Gober, M.D., et al. 2007. Novel Homozygous Frameshift Mutation of EVER1 Gene in an Epidermodysplasia Verruciformis Patient. *J Invest Dermatol* 127(4):817–820.

Lei, Y.-J., et al., 2007. HPV-2 Isolates from Patients with Huge Verruca Vulgaris Promoter Activities. *Intervirology* 50(5):353–360.

Wang, C. W. W., et al. 2007. Multiple Cutaneous Horns Overlying Varruca Vulgaris Induced by Human Papillomavirus Type 2: A Case Report. *Br J Dermatol* 156:760–762.

Wang, C., et al. 2007. Detection of HPV-2 and Identification of Novel Mutations by Whole Genome Sequencing from Biopsies of Two Patients with Multiple Cutaneous Horns. *J Clin Virol* 29:334–342.

Internet Resources

Indonesian Treeman, Department of Dermatology, University of Maryland, Dr. Anthony A. Gaspari, http://www.umm.edu/dermatology/treeman.htm

My Shocking Story—Half Man Half Tree: Discovery Channel, http://press.discovery.com/asia-pacific/dsc/programs/half-man-half-tree/

Chapter 2

A Good Deed Nearly Fatal

*This viral encounter represents a serious **zoonotic** viral disease. In countries like China, Vietnam, and Thailand, it is one of the most common fatal infectious diseases in humans. Worldwide, about 55,000 people die annually of this vaccine-preventable illness. Although people of any age are at risk, it often strikes children under the age of 15. Education and awareness programs must continue to prevent and control the spread of this disease.*

Fifteen year-old Jeanna Giese attended St. Mary Springs High School in Fond du Lac, Wisconsin. She was a straight-A student and star volleyball player.

"Do you see that?" whispered Jeanna.

"Shhhhhhhh," her brother nudged her.

They were attending a church service. A bat was flying around the ceiling and was headed straight toward the stained glass windows. It started flying lower and lower. A parishioner swatted at it and the bat dropped to the floor in the back of the church.

Jeanna loved animals and hated to see the bat injured and trapped inside the church. She went to the back of church and picked the bat up. She had never observed a bat so closely before. She stroked it and walked with it down the church steps.

"Ouch!" said Jeanna. "It bit me!" The bat was stuck hard to her left index finger. Jeanna threw the bat down and walked away. Later, she cleaned the wound with hydrogen peroxide and continued with her normal daily activities. At that moment, a bat bite didn't seem to be anything to get nervous about.

About three weeks later, Jeanna's left arm began to tingle.

"Maybe it's a pinched nerve?" said her mother. "Maybe you hurt yourself playing volleyball?"

Another week passed. Jeanna played in the last volleyball game of the season against a rival team. The game became surreal to her as she started seeing double. She jumped up to spike the volleyball and wondered if someone was playing tricks on her. Which ball should she spike?

The next day, Jeanna woke up exhausted. She felt achy and nauseous. She vomited several times. Her left arm started to jerk. She felt unsteady. Her speech was slurred. Her parents, John and Ann Giese, were concerned and took her to the emergency room where doctors quickly referred her to a neurologist. The neurologist assigned a battery of tests, including **magnetic resonance imaging** (MRI) of her brain. "The MRI showed no signs of brain damage or abnormalities," said the neurologist. He looked puzzled. There was nothing in her medical history to help make sense of her condition. All tests were coming up negative. She was not responding to any treatment.

Jeanna's symptoms progressed further. A fever set in. Her eyes moved involuntarily. Her motor skills were nearly nonexistent. Jeanna was admitted to the hospital with a 102° F fever.

Her mother kept retracing Jeanna's medical history and recent events in her head. She started writing in a diary. Her mind was racing. Then she remembered the bat incident in the church a month earlier. Could that be related to Jeanna's illness?

She immediately told the doctor what she remembered. "Jeanna was bit by a bat after mass about a month ago," she said cautiously.

The doctor's face turned white. "Did you seek medical attention?" he responded.

"No, at the time, we didn't think it was anything to worry about," replied Jeanna's father.

The doctor knew immediately that there was a good possibility that Jeanna could be suffering from rabies. Vaccination for rabies could be given after exposure, but only before symptoms started to manifest. All the textbooks stated that once symptoms began, no one survived rabies. The illness was always fatal. Treatment consisted only of sedation, controlling pain, and making sure the patient was restrained until death.

"I'm going to refer Jeanna to Dr. Rodney Willoughby at Children's Hospital," said the neurologist. "Dr. Willoughby is a pediatric infectious disease expert."

After Jeanna was transferred to Children's Hospital, her condition continued to deteriorate. By the second day there, she began frothing from her mouth (**hypersalivating**) and had difficulty swallowing. She was put on a respirator.

Dr. Willoughby was briefed about Jeanna's condition. He immediately ordered samples of serum, cerebrospinal fluid (CSF), saliva, and a skin biopsy from the nape of Jeanna's neck. All of the samples were sent to the Centers for Disease Control and Prevention (CDC) for laboratory diagnosis.

The tests came back with a positive rabies diagnosis. Rabies-specific antibodies were detected in Jeanna's CFS and serum samples. However, no rabies virus particles, antigens, or nucleic acid were detected from the skin biopsies or saliva samples. Perhaps the virus had not fully invaded her body yet?

Nonetheless, Dr. Willoughby's heart sunk. He had never seen a case of rabies, but he knew it was a death sentence. Nonetheless, he wanted to try to do something. Being proactive was better than watching Jeanna die without trying everything he could to help her. There was still much that was not understood about rabies virus infection.

Dr. Willoughby began researching the medical literature. He didn't have time to read papers thoroughly. He scanned many abstracts on rabies research and clinical case reports. Then he stumbled upon an obscure paper that mentioned sedation of rabies patients. He started to rationalize that it may be possible to stop the virus in its tracks. If he could stop Jeanna's brain from functioning, it might give her body enough time for her immune system to catch up and clear the virus with a vigorous antibody response and still save her brain. Dr. Willoughby and a team of doctors designed a radical treatment that would later become known as the "Milwaukee protocol."

Dr. Willoughby counseled Jeanna's parents. "Jeanna has rabies." Normally, rabies is 100% fatal. In fact, no one in the world has ever survived rabies once the symptoms start to show. We can sedate Jeanna until she dies either at the hospital or in hospice care at home."

The tears were streaming down their faces as they listened. They were intent on the doctor's words, but at times the thought of losing their daughter was overwhelming.

Then Dr. Willoughby said, "There is an option to try an aggressive approach which is an *untested strategy*. The idea behind this strategy is to sedate or turn Jeanna's brain off until her immune system can clear the virus from her body. We would give her a cocktail of drugs, which would put her into a coma intentionally, at which point we would give her an antiviral drug that might help knock down the virus. She would have the best supportive intensive care we can offer. We would monitor her brain activity and rabies-specific **antibody** response. We would then wean her from the meds, once we thought she had cleared the virus. There are, of course, many risks with this strategy."

Dr. Willoughby paused. He knew Jeanna's heart might fail with this aggressive approach. Jeanna could also sustain brain damage or be in a "locked in" state, not being able to move or communicate but aware of her surroundings.

He continued, "Even if Jeanna survives, she may be brain-damaged for life."

Her parents knew that this option was Jenna's only hope. The other alternative was death. This gave them hope. They also knew that if this option saved Jeanna, the strategy could save others. They decided the risks were worth it.

The hospital staff gave Jeanna ketamine and midazolam to induce the **coma**. She was also given phenobarbital to control seizures. After more consultation with CDC experts, the **antivirals** ribavirin and amantadine were administered after Jeanna was comatose. To avoid contracting rabies, all doctors, nurses, visiting relatives, and friends wore full isolation gear in Jeanna's presence: gown, mask, gloves, and a plastic face shield. While Jeanna lay in a coma, friends and relatives in yellow isolation gowns stood in a prayer circle around her.

After 6 days, the coma-inducing drugs were weaned.

Jeanna's eyes opened on the 10th day. Dr. Willoughby agonized over the outcome. Would Jeanna be severely brain damaged? What kinds of damage had the virus done? Could her mind respond?

Ann took her mask off. Jeanna's eyes tracked her mother. This was the first sign that Jeanna might have a hope of recovering.

Jeanna was to recover, although she had a rough road ahead of her. Her brain was "unwired." She couldn't talk or hold her head up. Her arms jerked. She couldn't relax her jaw and suffered from many uncontrolled movements. At the beginning of recovery, her attention span was 10 to 30 seconds. She underwent intensive rehabilitation.

After 11 weeks in the hospital, Jeanna was able to go home, where the long road to recovery continued. During an interview, her father said that "When Jeanna came home, her emotions were messed up. She was in a wheelchair, barely talking. She had to learn to crawl and then walk all over again."

Each day, improvements were made. Everyone was wondering how much of her former self would return as she revisited steps of childhood and learned old skills anew.

Jeanna Giese has received worldwide media attention as being the first person to ever survive rabies without post-exposure prophylaxis.

Update

Dr. Willoughby co-authored papers about Jeanna's survival after treatment of rabies through administration of a medically induced coma. At least 33 attempts using the Milwaukee protocol have been made with limited success. At the time of this writing, Jeanne Giese and five others have survived rabies using the Milwaukee protocol (or a very similar variation of the protocol). Some people suggest that these individuals survived because their bodies were able to mount an abnormally high antibody response to the rabies virus or that the strain of rabies with which these survivors infected was a weaker strain, allowing their immune systems to clear the virus. Either way, it has been seven years since a rabid bat bit Jeanna Giese. She has gained back much of her typical self. In the Spring of 2011, she graduated as a biology major from Lakeland College. Dr. Willoughby attended her graduation. She still suffers from slurred speech and some fine motor problems. The Giese family has been involved in rabies awareness projects.

Questions to Consider

1. What is postexposure prophylaxis for rabies?

2. Why can rabies vaccine prevent illness in someone already infected with the rabies virus?

3. Why is a bite on the foot from a rabies-infected animal somewhat less urgent than a bite on the face?

4. What should you do if you come in contact with a bat?

5. What other animals contract rabies in the wild and can transmit the virus to humans?

6. What are the symptoms of furious rabies?

7. What are the symptoms of paralytic rabies?

8. How is rabies diagnosed in an animal?

9. Rabies used to be called hydrophobia. Why?

10. How long can rabies be present in an animal or person before signs of rabies begin to show?

References

Holzamann-Pazgal, G., et al. 2010. Presumptive Abortive Human Rabies—Texas, 2009. *MMWR* 59(7):185–190.

Hu, W. T., et al. 2007. Long-Term Follow-Up After Treatment of Rabies by Induction of Coma. *NEJM* 352(24):2508–2514.

Hunter, M., et al. 2010. Immunological Correlates in Human Rabies Treated with Therapeutic Coma. *J Med Virol* 82: 1255–1265.

Kuzmin, M. D., et al. 2007. Human Rabies—Indiana and California, 2006. *MMWR* 56(15):361–365.

McDermid, R. C., et al. 2008. Human Rabies Encephalitis Following Bat Exposure: Failure of Therapeutic Coma. *CMAJ* 178(5):557–561.

Rubin, J., et al. 2009. Applying the Milwaukee Protocol to Treat Canine Rabies in Equatorial Guinea. *Scand J Infect Dis* 41(5):372–380.

Van Theil, P.-P. A. M., et al. 2009. Fatal Human Rabies Due to Duvenhage Virus from Bat in Kenya: Failure of Treatment with Coma-Induction, Ketamine, and Antiviral Drugs. *PLoS Negl Trop Dis* 3(7):e428.

Wiedeman, J., et al. 2012. Recovery of a Patient from Clinical Rabies—California, 2011. *MMWR* 61(04):61–65.

Wilde, H., Hemachudha, T., Jackson, A. C. 2008. Viewpoint: Management of Human Rabies. *Trans Royal Soc Trop Med Hyg* 102(10):979–982.

Willoughby, R. E., et al. 2005. Survival after Treatment of Rabies with Coma Induction. *NEJM* 252(24):2508–2514.

Willoughby, R. E. Jr. A Cure for Rabies? *Sci Amer* April, 2007 297(4):88–95.

Willoughby, R. E., et al. 2008. Generalized Cranial Artery Spasm in Human Rabies. *Dev Biolog (Basel)* 131:367–375.

Willoughby, R. E., et al. 2009. Tetrahydrobiopterin Deficiency in Human Rabies. *J Inher Metabol Dis* 32(1):65–72.

Internet Resources

Rabies CDC home page, http://www.cdc.gov/RABIES/

USDA—APHIS—Wildlife Rabies Management, http://www.aphis.usda.gov/wildlife_damage/oral_rabies/index.shtml

Rabies Awareness home page, http://www.rabiesawareness.com/

Medical College of Wisconsin Rabies, http://www.mcw.edu/display/docid18259.htm

Rabies Registry home page, http://www.chw.org/display/PPF/DocID/33223/router.asp

Global Alliance for Rabies Control, http://rabiescontrol.net/EN/Media-Center/Video-and-Audio.html

Jeanna Giese Rabies Survivor home page, http://site.jeannagiese.com/Home_Page.html

Chapter 3

A Case of the Trots

The "stomach flu" can be caused by a variety of infectious agents. People in crowded conditions such as schools, hospitals, and cruise ship liners are vulnerable to gastrointestinal illnesses caused by highly contagious infectious agents.

Microbiologists Sally and Kevin Goldstein were excited to go on their first cruise. They met during graduate school and fell in love quickly and married a year later. They were too busy to take a real honeymoon at the time. Now it was 10 years later. Both of them had finished their PhDs in microbiology and were working in the field. Sally was a research scientist for a pharmaceutical company, and Kevin supervised the microbiological quality control over Aunt Betsy's Cookie Dough Company. Kevin did most of the planning for the cruise.

"So what's the name of the ship we are going on again?" said Sally a week before the cruise date.

"The *Sea Dancer*," replied Kevin. "The *Sea Dancer* had the highest ship inspection scores posted on the Centers for Disease Control and Prevention (CDC) vessel sanitation program website," affirmed Kevin.

"Oh yeah, that's right," replied Sally. "Leave it to a microbiologist to know that!"

The week passed quickly. Soon they were sitting on one of the decks of the ship while sipping cocktails.

"Are you getting hungry, honey?" asked Kevin.

"I'm starving. This sea air is making me really hungry," replied Sally.

Kevin said with a smile, "All right, let's go for the buffet!"

"You know how I feel about buffets and smorgasbords!" replied Sally cynically. "They are a plate full of germs!"

"Come on sweetie, remember that the *Sea Dancer* had phenomenally high inspection scores. Can't you relax for a change and take the risk? You are always so critical of everything." Kevin snorted, "Microbiology has ruined your eating habits!"

Sally sensed Kevin's irritation. He was so good to her and usually humored her idiosyncrasies. This was supposed to be a vacation trip full of wonderful memories. She was willing to give in this once for Kevin.

"Oh, fine. I'll live dangerously." Sally knew the odds of getting some kind of gastrointestinal upset were low. After all, Kevin researched the cruise ship reports carefully and they had discussed this issue *ad nauseum*.

"Wow, this day will make history! Sally ate at a buffet! I am ecstatic!" Kevin's eyes beamed. He danced around her like a little kid in a candy store.

"Alright already, let's go!"

The number of meals per day and the quantity of food available for consumption on this cruise ship was almost obscene. Everything at the buffet looked delicious. There were exotic items such as Chilean sea bass, lobster Pad Thai, and duck watermelon salad. There were desserts like peanut butter pie and Norwegian chocolate mousse.

"I wonder how long this food has been sitting in the buffet?" said Sally.

"Don't start this again," reiterated Kevin. "Let's just enjoy the food and ambiance!"

"So be it," Sally replied grudgingly.

The two of them wolfed down all kinds of food. "I have to admit, this is the most heavenly food I have ever eaten," said Sally.

"I agree. You've been deprived," said Kevin with a grin. "It's about time you tried eating buffet food."

They chatted with the couple sitting next to them, Tara and Bob from Arkansas, about their jobs, their research, and how this was their much-delayed honeymoon. After their food binge, they waddled back to their rooms like overstuffed penguins.

"I'm so full!" said Sally.

"Well, this is our vacation honeymoon. We might as well enjoy it! Let's take a siesta and let our intestinal bacteria digest this great food," replied Kevin.

They napped for nearly two hours.

"Let's check out what activities are going on." suggested Sally.

Kevin read the brochure left in their room. "Ballroom dancing lessons, a spa, dining, dining, and more dining, swimming, bingo, board games and music...."

"How about swimming? I should float since I my stomach is so bloated from our food binge," said Kevin.

"Go for it, I'll tan on the side of the pool and read a book." said Sally.

Kevin accidently swallowed a huge mouthful of water after diving into the pool. "Well, that tasted good......... NOT!" said Kevin.

The day passed quickly. About 36 hours after their buffet binge, Kevin started getting abdominal pain and cramps. *This isn't good*, he thought. *It would be horrible if I had appendicitis on this trip.* He began to rationalize. *It's probably just all of those vegetables I've been eating. They are started to percolate in my guts.* Suddenly, he had an urgent need to go to the bathroom. He had watery, loose diarrhea. There was no way he was going to say anything to Sally. She would blame it on eating at the buffet immediately! She'd stop eating and would never go on a cruise with him again. Kevin came out of the bathroom, trying to hide his recent bout of intestinal dysfunction.

"Are you up for dance lessons?" said Sally.

"Oh honey, can't we do something more sedentary? We are supposed to be relaxing," whined Kevin.

"Hey, it's your turn to reciprocate. I ate at the buffet and I didn't want to. Now it's your turn to take dance lessons with me."

"Why do most men avoid ballroom dancing so much?" said Sally under her breath.

"Oh, ok. I can't argue with a scientist." Kevin uttered. He was starting to feel weak, nauseated, and had more stomach pains. He tried to hide how he was feeling, but it was becoming more difficult. As soon as they arrived in the cruise ship dance studio, he said, "Nature calls, be right back." He was in the bathroom nearly 30 minutes with vomiting and diarrhea.

"Are you alright? You look peaked." Sally looked concerned.

"Oh, it's just a little upset stomach." Kevin said. "It's nothing to worry about. Maybe I'm seasick....Sorry, honey... be right back." Kevin raced to the bathroom. He was truly uncomfortable, and the bathroom was occupied. He raced to find another bathroom on the ship. The next one was occupied! He searched for another that was also occupied. He finally found a free bathroom. After he finished, Kevin called Sally with his cell phone from the bathroom.

"I think I picked up some bug." Kevin said.

"I went back to the room," said Sally. "I'm not feeling well, either. "I feel like I have a fever, and it's coming out of both ends!" Sally sounded panicky.

"Let's not get graphic!" said Kevin. He was squeamish.

"I told you we shouldn't have eaten the buffet! Now we both have **gastroenteritis**! This is supposed to be a pleasurable trip!" wailed Sally.

"We are both microbiologists," said Kevin. "Let's be rational. We are both sick at the same time so we can probably rule out things like appendicitis. I really doubt we are seasick. It's probably gastroenteritis caused by a parasite, bacteria, or virus" he said. Kevin quickly made his way back to the room between bouts of diarrhea.

They both realized that they could be infected with a gamut of different infectious agents: parasites such as *Giardia* or *Cryptosporidium*, or bacteria such as *E. coli* or *Salmonella*, or even a norovirus. Kevin and Sally had eaten at the buffet and Kevin had swallowed pool water. Some of these agents are resistant to chlorine, so they could even be in the ship's drinking water.

As Kevin entered the room, the captain of the ship came on the intercom and made a cruise-wide announcement. "I have just learned that some passengers on this cruise are experiencing a stomach flu-like illness. If you have symptoms such as fever, diarrhea, or vomiting, do not use the public restrooms. Please stay quarantined in your staterooms until you are no longer have symptoms," said the captain. "Please report your illness to the ship's infirmary so that we can get a handle on how many passengers and crew are sick. Room service will be provided to affected passengers and crew. Every effort will be made to make everyone as comfortable as possible. If you need help, free healthcare is available on board. All passengers need to be extra vigilant and wash their hands before entering catering facilities. Enhanced sanitation protocols have already been implemented to help minimize transmission to other passengers. Additional hand sanitizers will be delivered to all staterooms. I will be regularly updating the passengers and crew on this situation. We are confident that this illness will be quickly controlled and will be eradicated in a few days time."

The calls started pouring in. Twenty of the 520 crew reported being sick. Over 100 of the ship's 1300 passengers called in their symptoms as well. This was 6% of the people on board the ship. (Cruise ships are required to report to the CDC when 2% of the people aboard exhibit gastrointestinal illness, defined as three or more episodes of loose stools in a 24-hour period or after vomiting with either loose stools, abdominal cramps, headaches, muscle aches, or fever.)

Because of the illness on board, the cruise liner was refused entry to port. The ship anchored outside of a different port, and CDC inspectors were brought in via a small boat to gather samples for testing. Stool samples were collected from sick passengers and crew. Pool, spa, drinking water, and food samples were also collected for analysis. The inspectors reviewed the ship's medical logs.

While sitting in her stateroom, still feeling miserable, Sally heard her cell phone ring. "Sally, this is Tara. Bob and I were sitting next to you two in the buffet room 2 days ago. I know you two are microbiologists, and I just want to know: Are you two doing some kind of germ experiment on this ship?"

"No, of course not Tara!" exclaimed Sally.

"What do you two think this is, seeing as how you guys are the experts?" asked Tara.

Sally answered cautiously, "A number of germs could be causing this. It could be a parasite, a bacterium, or a virus. I am beginning to think it might be a virus."

"What do we do?" Tara sounded concerned.

"Drink plenty of liquids. Wash your hands. Stay in your cabin, and keep listening for announcement updates."

"Aren't there vaccines for all these diarrhea bugs?" said Tara.

"Unfortunately not," said Sally.

"This ship is starting to smell like bleach, chlorine, and Purell!" yelped Tara.

"I assure you," said Sally, "this is the best way to decontaminate the ship and to prevent spread to all the other passengers. These bugs can go through a ship like wild fire. The tight quarters of a ship provide ideal conditions for contagious agents. If I'm right and this is a norovirus outbreak, these viruses are common, hardy, highly contagious, and hard to track. Given the close conditions, periodic outbreaks on ships may be inevitable," said Sally. "I gotta go Tara. Literally," said Sally, and she darted to the bathroom.

Just then, the captain made another announcement. "The CDC has confirmed a norovirus outbreak on the ship. The virus can cause diarrhea, stomach pain, and vomiting within 24 to 48 hours of infection. It is spread through food, water, and close contact with infected people or objects that infected people have touched. Tracing the norovirus to a common source can takes weeks to months. Thus, as of yet, we have not yet identified the source of the contagion. At this time, the number of passengers calling the infirmary has decreased. We are continuing the same practices of control, including cabin isolation of sick passengers, and decontamination. The *Sea Dancer* cruise line apologizes for this inconvenience."

"I guess even the viruses wanted to come to our belated honeymoon cruise," said Kevin.

Sally smiled and said, "Well, let's not take any samples back to the lab."

The cruise ended early, and passengers received a 50% refund because the cruise length was shortened. The ship was taken out of service to be disinfected.

Update

The CDC has been involved in surveillance for enteric diseases aboard passenger cruise ships since 1975. Since 2002 there has been an increase in norovirus-associated gastroenteritis on cruise ships. Inspections have led to improvements in ship construction, improved water sanitation, and limiting the amount of time that foods are on display rather than holding food at specific temperatures to avoid foodborne illness. Violations continue to decrease as the annual inspections go on.

Questions to Consider

1. Why are norovirus outbreaks so common?

2. Norovirus outbreaks don't just occur on sea-going vessels. Surf the Internet and list norovirus outbreaks not associated with cruise ships.

3. What preventable measures should one take to avoid a norovirus infection?

4. List the ways in which people become infected with norovirus.

5. When do norovirus symptoms usually appear?

References

Boxman, I. L., et al. 2009. Environmental Swabs as a Tool in Norovirus Outbreak Investigation, Including Outbreaks on Cruise Ships. *J Food Prot* 7(1):111–119.

Carling, P. C., et al. 2009. Cruise Ship Environment Hygiene and the Risk of Norovirus Infection Outbreaks: An Objective Assessment of 56 Vessels over 3 Years *CID* 49:1312–1317.

Cramer, E. H., Blanton, C. J., Otto, C. 2008. Shipshape: Sanitation Inspections on Cruise Ships 1990–2005, Vessel Sanitation Program, Centers for Disease Control and Prevention. *J Environ Health* 70(7):15–21.

Domenech-Sanchez A., et al. 2011. Unmanageable Norovirus Outbreak in a Single Resort Located in the Dominican Republic. *Clin Microbiol Infect* 17(6):952–954.

Flemmer, M., Oldfield, E. C. 3rd. 2003. The Agony and Ecstasy. *Am J Gastroenterol* 98(9):2098–2099.

Hall, A. J., et al. 2011. Updated Norovirus Outbreak Management and Disease Prevention Guidelines. *MMWR* 60(RR03):1–15.

Karst, S. M. 2010. Pathogenesis of Noroviruses, Emerging RNA Viruses. *Viruses* 2:748–781.

Neri, A. J., et al. 2008. Passenger Behavior During Norovirus Outbreaks on Cruise Ships. *J Travel Med* 15(3):172–176.

Rooney, R. M., et al. 2004. A Review of Outbreaks of Foodborne Disease Associated with Passenger Ships: Evidence for Risk Management. *PHR* 119(4):427–434.

Vivancos, R., et al. 2010. Norovirus Outbreak in a Cruise Ship Sailing around the British Isles: Investigation and Multi-Agency Management of an International Outbreak. *J Infect* 60(6):478–485.

Internet Resources

CDC Norovirus home page, http://www.cdc.gov/ncidod/dvrd/revb/gastro/norovirus.htm

CDC Vessel Sanitation Program home page, http://www.cdc.gov/nceh/vsp/default.htm

Chapter 4

Mysterious Transplant Complications

The diagnosis in this instance required tireless effort on behalf of the investigators involved. The mysterious viral encounter was solved through the application of a new technology that is paving the way to reopen past medical mysteries or to identify newly emerging pathogens.

Alexander Crowe was a 57-year-old banker living in Victoria, Australia. He recently retired and was planning several trips around the world. He loved to travel and was excited to spend 3 months traveling the rural areas of the former Yugoslavia. It was the trip of a lifetime for him. He traveled the countryside with delight and wonder. The pine forests and meadows of wild mountain flowers enchanted him. He visited many established tourist farms in the country. He loved the home cooked hearty meals. The time passed quickly, and soon his vacation was over and he was on a plane, returning to Australia.

Ten days after his trip he was back on the courts playing tennis with his buddies. During a rally, Alexander suddenly started feeling weak and nauseated. He developed a tremor in his left hand. His head was pounding. "Are you ok?" asked Ralph, his tennis partner.

As Alexander began to speak the word "No," he collapsed.

Within minutes of a 911 call, Alexander was taken by ambulance to a local hospital emergency room. When he arrived, he was unconscious. "This man looks like he is in tremendous physical shape." The attending doctor looked perplexed. His collapse wasn't likely the result of a sport-related event.

Just then, Alexander started to have a seizure. Emergency room doctors immediately administered anticonvulsants. Hospital personnel were frantically trying to make contact with his relatives to get Mr. Crowe's medical history.

"We need a head **CT scan** NOW!" yelled Dr. Keech. "My gut instinct tells me this guy has a brain aneurysm that burst. If I am right, his brain is hemorrhaging and he'll need to be prepped for surgery immediately."

At that moment, Alexander went into cardiac arrest. Doctors were unsuccessful in their cardiopulmonary resuscitation (CPR) attempts. Time of death was announced at 3:02 PM, and cause of death was attributed to a cerebral hemorrhage or what may be referred to as a *stroke.* Hospital staff noted that his driver's license indicated he was an organ donor.

Dr. Keech said, "Call the family and confirm that we can prepare his organs for donation."

Three women in Australia on a liver or kidney transplant waiting list were called about a matching donor being available. All of them were in life-threatening need of an organ and were ecstatic to learn that they soon may have a transplant that would save their lives. Sixty-three-year-old Anna Miller and 44-year-old Marie Gilmore were prepped for kidney transplants while 64-year-old Anna Anderson was prepped for a liver transplant.

The surgical procedures were uneventful, and all of the transplant recipients emerged from the surgeries with a new hope for their lives. Within weeks of their transplants, however, all three recipients developed a similar fever with varying degrees of encephalopathy. Each one of them died 4 to 6 weeks after receiving the transplant.

Doctors were puzzled. The deaths were not caused by rejection of the transplanted organs, which is the usual concern or complication in any transplant case. To help track down the causes of these deaths, the women's tissues were tested using a method called **polymerase chain reaction** (PCR), looking for genomic sequences of parasitic, bacterial, and viral pathogens. The PCR assays screened for the bacterial pathogens *Mycobacterium tuberculosis* and *Mycoplasma pneumoniae* and the parasite *Toxoplasma gondii.* All these results were negative. PCR was then used to screen for the following viral pathogens: herpesviruses 1–8, lyssavirus,

influenza A and B viruses, respiratory syncytial virus, picor-
naviruses, adenoviruses, human parainfluenza virus, flaviviruses, alphaviruses, hantavirus, polyomavirus, Crimean-
Congo hemorrhagic fever virus, and Rift Valley Fever virus.
Again, all results were negative. PCR was unable to detect the
genetic sequences of any known pathogens.

Since the PCR method, which identifies the cause of disease based on knowing the genetic sequence of the pathogen involved, was unsuccessful, attempts were also made
to culture pathogens in the laboratory from patient tissues.
Serologic tests were performed to identify **IgM antibodies**
against a gamut of known pathogens that might indicate
a recent microbial infection. These tests too, turned up no
leads. **Microarray analysis** for a panel of infectious agents
was also negative. Doctors were stumped.

A team of medical researchers from the Victorian Infectious
Disease Reference Laboratory, the CDC, and Columbia
University persevered in the quest to unfold the cause of
this mysterious cluster of deaths. They used a new technique called **unbiased high-throughput sequencing** in an
attempt to identify a new pathogen that may have been
potentially transmitted through solid-organ transplantation.
The method involved isolating the RNA from the brain,
cerebrospinal fluid (CSF), **serum**, kidney, and liver of the
transplant recipients. The RNA was amplified using **reverse-
transcriptase PCR** (RT-PCR) using random primers. These
amplified DNA products were pooled and sequenced. The
sequences obtained were assembled into a set of sequence
data. Any sequences that matched human sequences in the
GenBank database were removed from consideration. Any
unknown or unique sequences were analyzed and compared
to sequences of pathogens in the **GenBank database**. It was a
tireless effort by this team of investigators.

After many hours of pouring over the data, "Eureka!" cried
Dr. Lipkin, a Columbia University professor of microbiology.
He was conversing with Dr. Catton at the Victoria Infectious
Diseases Reference Laboratory. "These novel sequences are
similar to two arenaviruses: lymphocytic choreomeningitis
virus (LCMV) and Kodoko virus."

"This may be an infection caused by a new strain of arenavirus!" said Dr. Catton. "There was a cluster of organ recipients that died of a LCMV infection in 2006! In that cluster of

cases, the donor was infected by a pet hamster that carried the virus."

"The next step is to backtrack. We need to reanalyze the tissue samples from the liver and kidney recipients to confirm this," replied Dr. Lipkin.

"This is so exciting!" enthused Dr. Catton.

Laboratory workers began to inoculate Vero E6 (monkey kidney epithelial) tissue culture cells with homogenates of fresh-frozen kidney or liver from the transplant recipients in an attempt to isolate the new strain of virus. **Immunofluorescent staining** of the inoculated cell culture cells probed with antibodies against arenaviruses and LCMV showed the presence of viral antigens. They examined these cells with electron microscopy. They observed particles consistent with the same morphological characteristics as other arenaviruses.

"It's there," said Dr. Du, one of the researchers. "We see viral particles!"

"All we have to do is a serologic analysis and to confirm the culture by RT-PCR," said his colleague, Dr. Tran.

"Results are in," said Dr. Zaki, yet another tireless researcher on this case. "We've been able to detect identical sequences of the new arenavirus RNA in 20 of 30 specimens collected from the three transplant recipients."

"We've also been able to detect virus-specific IgM antibodies in serum samples of all three transplant recipients as well," chimed in the final researcher, Dr. Goldsmith. "This indicates they've been recently infected with the new virus."

"We've got to make sure our investigation is thorough," said Dr. Lipkin, when the results were reported to him. "Let's test 100 of the archived serum or plasma samples we have from other solid organ transplant recipients who were not linked to this cluster but whose transplants were performed in the hospitals in Victoria during this same time period."

"We are already ahead of you," said Dr. Catton. "It's been done. None of the archived samples show any evidence of this new viral pathogen."

"What about the donor in this case?" inquired Dr. Lipkin.

Dr. Catton replied, "We didn't detect any viral RNA sequences to match the new arenavirus in the donor. But we did detect IgG and IgM antibodies in serum samples from the donor."

"So the donor had a recent infection and had **seroconverted**," said Dr. Lipkin.

"Where do you think he picked up this new arenavirus?" said Dr. Tran. He was listening intently to the conversations buzzing in the laboratory.

"The *Arenaviridae* are a family of viruses whose members are generally associated with rodent-transmitted disease in humans," began Dr. Catton. Each virus usually is associated with a particular rodent host species in which it is maintained." Arenavirus infections are relatively common in humans in some areas of the world and can cause severe illnesses. I seem to recall that the donor had recently visited the former Yugoslavia, where he traveled in rural areas where rodents may be prevalent."

"I think high-throughput sequencing just solved our mystery!" said Dr. Lipkin. "This powerful tool may allow others to discover more new pathogens in order to solve new and remaining medical mysteries."

"What a super joint effort!" said Dr. Catton.

Update

High-throughput sequencing continues to allow researchers to search for unrecognized pathogens including the early detection of emerging strains of arenaviruses, noroviruses, and seasonal influenza A viruses. Early detection of these strains has received renewed attention.

Questions to Consider

1. What is a zoonotic infection?

2. How are arenaviruses transmitted to humans?

3. Why was this such a difficult medical mystery to solve?

4. What precautions can be taken to avoid an arenavirus infection?

5. What methods were employed to confirm that this new virus came from the transplant donor?

References

Allstair, C., Hall, N. 2008. High-Throughput Sequencing Rapidly Connects Microbial Phenotypes and Genotypes to Guide Metabolic Engineering. *Nature Biotechnol* 26:1248–1249.

Amman, B. R., et al. 2007. Pet Rodents and Fatal Lymphocytic Choriomeningitis in Transplant Patients. *Emerg Infect Dis* 13(5):719–725.

Fischer, S. A., et al. 2006. Transmission of Lymphocytic Choriomeningitis Virus by Organ Transplantation. *NEJM* 354:2235–2249.

Nakamura, S. 2009. Direct Metagenomic Detection of Viral Pathogens in Nasal and Fecal Specimens Using an Unbiased High-Throughput Sequencing Approach. PLoS *ONE* 4(1):e4219.

Palacios, G., et al. 2008. A New Arenavirus in a Cluster of Fatal Transplant-Associated Diseases. *NEJM* 358(10): 991–998.

Peters, C. J. 2006. Lymphocytic Choriomeningitis Virus—An Old Enemy up to New Tricks. *NEJM* 354:2208–2211.

Internet Resources

Arenaviruses CDC Special Pathogens Branch home page, http://www.cdc.gov/ncidod/dvrd/spb/mnpages /dispages/arena.htm

Chapter 5

Pass Me the DEET

Re-emerging diseases are ones that were on the decline in a given population but have increased due to changes in the environment and susceptible populations. The virus in this encounter is an imported cause of a disease that was once an old scourge in southern parts of the United States but then disappeared. Now it is making a deadly comeback.

"Ow! That hurt!" said 28-year-old Melissa Lopez as she swatted at a mosquito that bit her on the arm. "These mosquitoes are so annoying!"

Melissa Lopez, a biology student at UC Berkeley, was on vacation in Chacalapa, Mexico, with her boyfriend, Daniel Rodriguez. They had been saving their money for a couple of years to go to Mexico and were excited to be there. Daniel had been born in Mexico and lived there until he was 12, when his family moved to Seaside, California. He wanted to show Melissa where he had lived as a boy.

"What is it with these mosquitoes? I thought they only came around at night?" said Melissa. Daniel and Melissa were eating lunch outside at a restaurant.

Daniel said, "They sure are aggressive!"

"I can't believe we forgot the bug spray!" said Melissa. "I am starting to itch already."

"Well, they are not biting me!" said Daniel. "Maybe you have too much perfume on, and the mosquitoes are attracted to your scent?"

"Maybe these are killer mosquitoes coming up from South America—kind of like those killer bees?" asked Melissa, only half-kidding.

"I wonder if **global warming** has anything to do with all of these mosquitoes?" said Daniel. "I read this thing about the **Asian Tiger Mosquito**. It came to North America from Asia in a shipment of tires or something in 1985, and now it's in something like 36 states! As the climate gets warmer, the bugs can go into territories they could never survive in before."

"It makes sense that as temperatures get warmer, the range of certain mosquitoes and the duration they are able to circulate increases," said Melissa, switching into bio major mode. "I think the proper term is **climate change**."

Daniel smiled and said, "Or is it global climate change? Some people call this *global whatever* junk science," added Daniel.

"Seems pretty clear to me that something is happening," replied Melissa.

After lunch, the two found an over-the-counter insecticide spray in a pharmacy and were able to enjoy the rest of their trip to Mexico. The time flew quickly, and soon they were back in California working at the restaurant where they had met. (Daniel was a chef, and Melissa waitressed at the Muchos Habaneros restaurant, making money to put themselves through school.)

About 4 days after their return, Melissa started feeling sick. She tried to ignore it. The restaurant was short-staffed, and she needed the money. She had to go to work. She skipped her morning run, which was unusual behavior for Melissa, who was in excellent shape and ran at least 6 days of the week. She usually took great care of herself and was very health conscious, making sure to exercise often and eat healthy foods. Melissa rarely got sick.

"Hey, you don't look good sweetie," said Daniel.

"Yeah, maybe I am getting the flu?" replied Melissa. "I feel really hot and my head is throbbing. My whole body aches."

"Maybe you should go home and rest?" said Daniel. "You don't have to do this to yourself. You hardly ever take a day off. Go home. Pamper yourself, and you'll be good as new soon. I'll come over after my shift and take care of you."

"I'll be OK. I can focus on the customers and get through this," strained Melissa.

A few hours later, though, Melissa started to feel nauseated. She went to the bathroom and vomited. A myriad of thoughts were running through her mind as she was sick. Could she be pregnant? The timing was awful if she were. Or, did she pick up some bug in Mexico? She had heard about illnesses like "**traveler's diarrhea**" upon return from a visit to Mexico. She didn't have the gamut of flu-like symptoms that she would have had if it were influenza, like fever, sore throat, headache, dry cough, muscle aches, and pains. Even though her symptoms didn't fit, she couldn't figure out what else it could be, so she decided she probably had the flu after all.

After her trip to the bathroom, she asked her boss if she could take the remainder day off. If she had something that was contagious, she certainly didn't want to make the customers sick. The boss agreed and Melissa went home. Her "flu" dragged on for a few days. At one point her fever dropped after profuse sweating, but then the fever returned. She started to have severe body aches, fever, and chills throughout the week. She had intense abdominal pain.

Daniel was concerned. "I think you should get a check up," he said when she called him at work, obviously in distress. "I am coming over to get you."

Melissa agreed. She decided to get ready to go to a walk-in clinic. She had no appetite and felt very weak and unsteady. She went to the bathroom and noticed blood in her stool. She began to shower and collapsed.

Daniel arrived. He knocked on the door, but there was no answer. He pounded again and still there was no answer, so he used his key to enter her apartment. He found her slumped over in the shower. Her eyes were red. Her skin and face was flushed. She had a rash that covered her entire body except her face. The palms of her hands and soles of her feet were bright red and swollen. He couldn't believe what was happening! What was wrong with her? He called 911 and paramedics were there within 15 minutes. Daniel called Melissa's family members to let them know she was on her way to the hospital.

Paramedics attended to her in the ambulance on the way to the nearest emergency room. Her blood pressure and heart rate were low. She had a rapid, weak pulse. Once they arrived at the hospital, the attending emergency room physician observed her carefully. "She's in shock," said Dr. Carlson. "She appears extremely dehydrated. Let's get some **intravenous fluids therapy** now! She needs the works: saline, Ringer's, glucose; stand by on plasma."

Then he noticed her rash. Her body extremities were cool and clammy, but her trunk was warm. Her pulse was weak. There was a blueness or **cyanosis** around her mouth. Dr. Carlson placed an oxygen mask on her.

Once she was stabilized, Melissa was admitted into intensive care. Nursing staff immediately hooked her up to sophisticated electronic monitoring equipment. Other staff was on the phone ordering and gathering medical history information. Dr. Carlson ordered laboratory tests: **hematocrit** (a blood test that measures the percentage of red blood cells found in whole blood), serum electrolytes and blood gas studies, platelet counts, liver function tests, sera to screen for IgM antibodies against yellow fever virus, West Nile encephalitis virus, and a rapid test for leptospirosis. The staff monitored her vital signs every 10 minutes.

Dr. Carlson knew the rash was an important clue in determining what was wrong with Melissa. He noticed bleeding at the IV injection site. He suspected she was bleeding internally. Closely monitoring the vital signs and hematocrit levels were important in assessing the severity of hemorrhaging.

Melissa was not responding to treatment. She required whole blood, platelet, and plasma transfusions. Despite the best intensive care procedures, she died. The family was grief-stricken but agreed to an autopsy.

Doctors were dumbfounded by this case. They contacted the Centers for Disease Control and Prevention (CDC) regarding this case and provided them with the results of the autopsy, which indicated that there was bleeding found in the skin and subcutaneous tissues, the mucosa of the gastrointestinal tract, and in the heart and liver. Capillaries in the affected organ systems revealed bleeding. Lymphocyte tissue showed an increase in activity of the B-lymphocyte system. There were lesions in the liver. **Dengue virus antigen** was found predominantly in the liver, spleen, thymus, lymph node, and lung cells. The virus was isolated at autopsy from bone marrow, brain, heart, kidney, liver, lungs, lymph nodes, and the gastrointestinal tract. The virus was grown in mosquito-derived cell cultures. IgM specific antibodies against the dengue virus were present in serum. Melissa Lopez had died of **dengue hemorrhagic fever (DHF)**. Virus isolation is widely used as the gold standard for diagnosing DHF.

She had contracted the dengue virus through a day-biting *Aedes* mosquito while on vacation in Mexico. The discovery of what killed Melissa shocked her family as well as the physicians and public health officials involved in her case.

Dengue fever was common in North America until the late 18th century, after which increased sanitation and, once the vector was identified as mosquitoes of the *Aedes* genus, mosquito eradication efforts helped to decrease the probability of outbreaks. Since global climate change and other instances of ecological disruption have altered the range and duration of *Aedes* mosquitoes, however, instances of dengue fever and its more lethal counterpart dengue hemorrhagic fever, are on the rise in many areas of the world, including Mexico and along the United States–Mexico border.

Update

Each year, there are approximately 50 million dengue virus infections with about 500,000 hospitalizations around the world. It is the second most common cause (after malaria) of feverish symptoms for Western tourists returning from developing countries. DF cases are on the rise in Mexico (1,781 cases in 2000 to 30,000 cases in 2008). In 2000, DHF represented one in about 26 cases; in 2008, it was one in 4. Travel-associated DF and limited outbreaks do occur in the continental United States; most DF cases in US citizens occur as endemic transmission among residents in some of the US territories. Aedes aegypti *has reemerged in the Americas following a successful eradication campaign during the 1950s and 1960s (see* **Figure 2***). Dengue fever was placed on the US CDC* Morbidity and Mortality Weekly (MMWR) *list of "reportable diseases" January 22, 2010, requiring laboratories and heath-care providers to report all US cases of dengue to the CDC.*

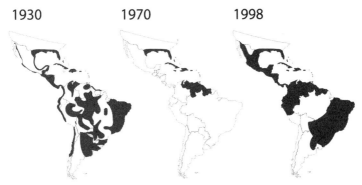

Figure 2. *Aedes aegyptii* in the Americas. (Data from CDC. Drawn by Brian Ledwell.)

Questions to Consider

1. Dengue is a disease caused by how many different sero-types of viruses?

2. How does DF differ from DHF?

3. In 1985, the Asian tiger mosquito was first found breeding in the United States. What states has it been found in to date?

4. Explain how climate change could turn the tide in favor of the mosquitoes?

5. List examples of man-made breeding sites for mosquitoes.

6. How can DF or DHF be prevented?

7. Who is at risk for DF?

8. Why was DF once called "breakbone fever"?

9. Can an individual get DF more than once? Explain your answer.

10. Why has it been difficult to develop a vaccine that prevents DF or DHF?

11. Why is vector control problematic?

References

Kyle, J. D., Harris, E. 2008. Global Spread and Persistence of Dengue. *Ann Rev Microbiol* 62:71–92.

Luce, R., et al. 2010. Travel-Associated Dengue Surveillance— United States 2006–2008. *MMWR* 59(23):715–719.

Streit, J. A., et al. 2011. Upward Trend in Dengue Incidence among Hospitalized Patients, United States. *Emerg Infect Dis* 17(5):914–916.

Trout, A., et al. 2010. Locally Acquired Dengue— Key West, Florida, 2009–2010. *MMWR* 59(19):577–581.

Internet Resources

Dengue Fever Fact Sheet, http://www.cdc.gov/ncidod /dvbid/dengue/

World Health Organization, 1997. *Dengue Haemorrhagic Fever: Diagnosis, Treatment, Prevention and Control*, 2nd edition, http://www.who.int/csr/resources/publications/dengue /Denguepublication/en/

United States Global Research Program, http://www .globalchange.gov/

Chapter 6

An Alternative Sushi

This viral encounter is about a virus that may reside in wild boar and deer populations. Humans may contract a viral illness through the consumption of raw pork and deer.

Seventeen-year-old Nicole Boury was thrilled about the opportunity to live with the Noto family in Japan for a summer. She had been best friends with Sarah Noto since the fifth grade when Sarah and her family moved to the United States. Sarah still had many relatives in Japan.

The two were inseparable. Nicole became enamored with Japanese customs and culture. She read voraciously about living on the islands of Japan. She daydreamed about seeing the snow-capped mountains of northern Hokkaido and the sandy shores of Okinawa. She was curious about Japanese cuisine and crafts. Through the years, Sarah taught Nicole to speak Japanese, and, when they turned 17, she was able to make arrangements for Nicole to stay with her Uncle Noto's family for 8 weeks in the summer. Nicole knew that this was unusual. It was an honor to be invited into someone's home in Japan, so she was particularly excited to go.

Sarah's aunt and uncle, Mai and Kai Noto, greeted Nicole at the Tokyo Narita International Airport. The first thing they did together was to eat at a Japanese restaurant. Nicole loved it! She had practiced using chopsticks at restaurants for months before the trip. She knew that Japanese food was one of the healthiest cuisines in the world: low in cholesterol, fat, and calories and high in fiber. At the entry to the restaurant, where the group removed their shoes, Nicole observed a glass display of plastic or wax replicas of the dishes at the restaurant. She assumed this was done to help foreign tourists who do not speak Japanese choose what they wanted to eat.

The waiters and waitresses were very polite and efficient. A waitress led them to a table and served them tea immediately. The Notos ordered a variety of dishes, and in less than 10 minutes, the food arrived at the table.

"*Itadakimasu*," said Mai with a smile.

"That means 'I humbly receive'!" said Nicole enthusiastically, ready to show off her Japanese. "In my family we say something similar: '*bon appétit.*'"

Nicole watched Kai dip sushi into soy sauce. She had read that many Japanese dishes were raw or only partially cooked: sushi, sashimi, tataki, namagimo, and shabu-shabu. She glanced around the room and noticed that people dining were not eating raw tuna sushi, but some other kind of uncooked meat.

"What's that?" asked Nicole.

"Looks like there is wild boar and deer available tonight," said Kai.

"It looks raw," whispered Nicole.

"It is," replied Kai. "After the **Mad Cow** scare in England, we are avoiding beef. There has also been a tuna shortage, and the prices of tuna are too high for some restaurant owners. As an alternative, we sometimes consume raw horse, pork, and deer sushi."

"Don't forget about the mercury poisoning scare in the 70s," said Mai. "Some customers refused to eat tuna because of the mercury that accumulates in tuna meat."

"Chefs are experimenting with more creative dishes like smoked duck with mayonnaise and crushed daikon radish with sea urchin," chimed in Kai.

"Look who just arrived at the restaurant!" Mai said to Kai. "Your sister Asa and brother Hirocki are here."

The dinner went by quickly. Nicole remembered she could slurp her ramen noodles. In fact, it was considered polite to do so! Before she knew it, it was time to leave.

"*Gochisosama deshita*" Nicole said to the waitress. "Thank you for the meal." She tried very hard to use the Japanese language that Sarah had taught her.

For the next few weeks, the Noto family took Nicole sightseeing around the greater Tokyo area and to several islands of Japan. They toured the Toyota, Nintendo, and Canon corporations as well.

When they visited Okinawa, it was the rainy season. Nicole was surprised by how forested Japan was. She learned that the majority of the land was used for agriculture, industry, or residential use but that a good deal of the land is too forested and mountainous to use. That is why Japan is one of the most densely populated countries in the world.

"There are wild boar, deer, serow, Japanese bear, and macaques present in the forest," Kai told Nicole.

"So is there a lot of hunting in Japan?" asked Nicole.

"Yes, there is," said Kai. "Tourists also have opportunities to hunt."

"What animals are hunted?" asked Nicole.

"There are Sika deer, razor-back wild boar, green and copper pheasants, turtle doves, partridges, quail, and black and brown bear…a great variety!" said Mai.

Just then, Mai's cell phone rang.

"What? Uncle Hirocki is in the hospital?" said Mai.

Everyone listened intently to the half-hour conversation with his sister about Hirocki. Mai explained, "Apparently Hirocki was complaining of a fever, nausea, and malaise, so he was taken to the Kasai City Hospital, where the doctors did some tests. His liver enzymes were elevated, and they concluded he had **acute hepatitis**," continued Mai. "The doctor said his **alanine aminotransferease** (ALT) level was at 2163 units per liter plasma and his **bilirubin** at 29 milligrams per deciliter serum."

"Normal ALT levels are in the range of 7 to 56 units per liter of serum," said Kai, who was a practicing nurse. "Normal bilirubin range is 0.1 to 1.2 milligrams per deciliter serum. Elevated levels of ALT are an indication that the liver is inflamed."

"What causes hepatitis?" asked Nicole.

"The doctors tested him for a number of viruses that can cause hepatitis. They tested him for hepatitis A, hepatitis B, and hepatitis C viruses, but he turned up negative," said Mai. "Some pharmaceuticals can also cause liver problems, but Hirocki doesn't take any medications, not even Tylenol. So far, the doctors are puzzled."

"Can we visit him?" asked Nicole. "The last time I saw him was at the restaurant when I first arrived in Japan."

The next day, they went to the hospital to visit Hirocki. Nicole waited outside of his room while Mai and Kai went in. She listened and observed the hospital staff. She found the conversations among the doctors fascinating.

Nicole overheard a doctor in a white coat talking. She thought his name was Dr. Tei. She tried to read the name badges on their white coats.

"There is a researcher here today to gather specimens from us on this mysterious hepatitis case. His name is Dr. Takahashi. He will run some virological analyses, and Dr. Mishiro will analyze and interpret his results."

Dr. Tei walked up to the nurse's station. "Are the serum samples read for Dr. Takahashi?" he asked.

"Yes," replied the nurse.

Then Nicole saw Dr. Tei conversing with Dr. Takashashi. "We will test this patient's serum for antibodies against hepatitis E virus. We also have more samples from members of the same family and another family," said Dr. Tei, "for comparison."

"More cases have trickled in," said Dr. Kitajima, who had joined the team of doctors. "The first patient, Mr. Hirocki Noto, traveled to China recently," said Dr. Tei. "We initially thought that the virus was imported from China and spread within the family. But then we took patient histories and were surprised to hear that both the Noto family and other patients presenting similar symptoms had eaten raw Sika deer sashimi or sushi three times during the past seven weeks. This makes the possibility of hepatitis E infection much more likely, but we still need to confirm it."

Dr. Mishiro joined the conversation, "This is a very interesting case. There are also reports that hepatitis virus E can be transmitted among people who eat undercooked wild boar," said Dr. Mishiro.

"I have read about some studies in progress that have detected hepatitis virus E in wild boars and deer of rural regions in Germany and eastern China as well," said Dr. Takashashi.

"Some of the patients froze the left-over portions of Sika deer to eat in the future," added Dr. Kitajima. "They gave us the frozen raw deer meat for testing."

"I will do **RT-PCR** on these samples and see if we can detect hepatitis E virus RNA," said Dr. Takahashi.

Nicole was fascinated by this conversation. Now it was time for her to visit Sarah's Uncle Hirocki.

A couple of weeks passed and Uncle Hirocki was able to go home. He visited Kai and Mai Noto's family during Nicole's last week in Japan.

"I guess the deer sushi made me sick," sighed Hirocki. "The doctor said the leftover uncooked deer we froze tested positive for the hepatitis E virus. Thank goodness none of you ate any."

"Oh deer!" said Mai.

Update

Hepatitis E virus is usually spread by drinking water contaminated with feces containing the virus. There is no specific treatment for the disease and it usually resolves on its own. Recent studies have proved that hepatitis E virus is a zoonotic virus and that domestic swine, wild boars, and wild deer are reservoirs of the virus in nature. There are at least four different genotypes or strains of the virus. Genotype 1 is found in developing countries of Asia and Africa; genotype 2 has been found in Mexico and Africa; genotype 3 is widely distributed throughout the world and has been isolated from domestic pigs in the United States, Europe, and Japan. Genotype 4 is found mainly in Asian countries. Other animals such as cows, sheep, camels, horses, dogs, cats, rats, and mongoose are also susceptible to hepatitis E virus infection and may be reservoirs of the virus.

Questions to Consider

1. How do the clinical features of hepatitis A, B, C, and D infection compare to hepatitis E virus infection?

2. List hepatitis E virus mode(s) of transmission.

3. How can hepatitis E virus infection be prevented?

4. What are the risk factors for hepatitis E infection?

5. How do the hepatitis viruses differ in terms of their structure and replication strategies?

6. Where is hepatitis E infection the biggest problem and why?

7. What is the incubation period for hepatitis E virus infection?

8. What is bilirubin and its association with hepatitis?

9. List all of the hepatitis viruses and whether or not vaccines are available to prevent disease caused by them.

10. Which hepatitis viruses cause chronic infections?

References

Ijaz, S., et al. 2005. Non-Travel-Associated Hepatitis E in England and Wales: Demographic, Clinical, and Molecular Epidemiological Characteristics. *J Infect Dis* 192:1166–1172.

Kazuaki, T., et al. 2009. Virulent Strain of Hepatitis E Virus Genotype 3, Japan. *Emerg Infect Dis* 15(5):704–709.

Lewis, H. C., et al. 2008. Hepatitis E in England and Wales. *Emerg Infect Dis* 14(1):165–166.

Mashuda, J.-I., et al. 2005. Acute Hepatitis E of a Man Who Consumed Wild Boar Meat Prior to the Onset of Illness in Nagasaki, Japan. *Hepatol Res* 31:178–183.

Meng, X. J. 2010. Hepatitis E Virus: Animal Reservoirs and Zoonotic Risk. *Vet Microbiol* 140(3–4):256.

Mushahwar, I. K. 2008. Hepatitis E Virus: Molecular Virology, Clinical Features, Diagnosis, Transmission, Epidemiology, and Prevention. *J Med Virol* 80:646–658.

Schielke, A., et al. 2009. Detection of Hepatitis E Virus in Wild Boars of Rural and Urban Regions in Germany and Whole Genome Characterization of an Endemic Strain. *Virol J* 6:58 doi;10.1186/1743-422X-6-58.

Tei, S., et al. 2003. Zoonotic Transmission of Hepatitis E Virus from Deer to Human Beings. *The Lancet* 362(938): 371–373.

Zheng, Y., et al. 2006. Swine as a Principal Reservoir of Hepatitis E Virus That Infects Humans in Eastern China. *J Infect Dis* 193:1643–1649.

Internet Resources

CDC Viral Hepatitis, http://www.cdc.gov/hepatitis/

WHO Hepatitis E fact sheet, http://www.who.int /mediacentre/factsheets/fs280/en/

Chapter 7

Frog Killer

Nearly a third of the world's frog populations are in danger of becoming extinct. This serious decline is due to a number of factors—habitat destruction, invasive species, climate change, and so on. But another cause of widespread frog mortality may not be so obvious—it may lie in the world of the microbes. This viral encounter addresses the mysterious disappearance of northern leopard frogs, which were once common in many parts of the United States and Canada.

Darin Peterson and Mike Robertson are herpetologists at Idaho State University studying the ecology and conservation biology of amphibians and reptiles. They love being in the field doing research on amphibians. One day, as they were hiking through the wetlands of Wyoming, Darin turned to Mike and said, "Do you hear that?"

"Hear what?" said Mike.

"Nothing. No croaking frogs," said Darin. "When I was a kid in the '70s, I used to catch northern leopard frogs (**Figure 3**) by the dozen. I could hear them croaking, and when I got up near the pond, I could see them jumping in."

"Life has certainly changed for these frogs," said Mike. "They just seem to be disappearing."

Some scientists blame the large decreases in populations of frogs and other amphibians on climate change or an increase in ultraviolet radiation levels due to ozone depletion, as frogs have very sensitive skin. In truth, their disappearance is complicated. It could be linked to environmental issues, disease, or habitat changes. In Darin and Mike's area, it could have to do with damming and channeling of the Snake River because lots of wetlands were drained in the process, decreasing the

Figure 3. The northern leopard frog. (Drawn by Brian Ledwell.)

frog population's usable habitat. Nearby cattle using the wetlands for grazing might put stress on the tadpole populations by polluting the water with their waste, reducing water quality. Drought can also cause physical stress on the animals. Such stress could compromise the frogs' immune systems and make them more susceptible to infection. Diseases caused by fungi, bacteria, or viruses can all kill frogs when they become susceptible.

Frog populations can also be harmed if their food chain is disrupted. Insecticides such as malathion are used to treat ponds to get rid of mosquitoes that carry human diseases, like West Nile encephalitis. This is a good thing for human health, but it cuts off food supply to tadpoles, so they are unable to live long enough to produce offspring. Some frogs don't breed until they are three or four years old and cannot reach breeding age if water quality and food sources are too low. Other research by scientists such as Rick Relaya suggests that the herbicide Roundup® is extremely lethal to frogs and other amphibians.

"I think the decline of frog populations is far more complicated than most people think," considered Darin. "It is rare when a species declines for one reason. There are frogs disappearing in ecologically pristine areas like the montane tropical rainforests of Australia and Central America."

"You're right," said Mike. "There's no human impact from agriculture or development, no deforestation or much pollution in those areas. I think it's more likely an emerging disease like chytridiomycosis or ranaviral disease that causes frog populations to fluctuate in areas like that," said Mike.

Chytridiomycosis is a fungal disease that was first discovered in Queensland, Australia in 1993, though it has probably been in Australia since the 1970s. It is also found in southern Africa; a group of researchers studied over 600 archival specimens collected in southern Africa from 1879 to 1999 and found a case of chytridiomycosis in an archived *Xenopus laevis* frog sample from 1938, the earliest known to date. South Africa is the proposed origin of the chytrid fungus (*Batrachochytrium dendrobatidis*), which is assumed to have been disseminated to frogs in other locations of the world through the international trade of frogs. Tadpoles and frogs are shipped around the world for commercial sale, for pets, scientific research, and even food. These frogs may be carrying the chytrid fungus but have no symptoms. If they are later released into their new environment, they can spread the fungus to other frogs.

Chytridiomycosis is now thought to be responsible for the decrease in populations of amphibians in North and South America, Europe, and even some areas of the Caribbean, in addition to Africa and Australia. The fungus first infects the skin, causing a reddening of the underside of the body and roughening and shedding of the skin. Other symptoms include convulsions of the hind legs, lethargy, and other abnormal behaviors.

Ranaviruses are a member of the Iridoviridae family of viruses. Like chytridiomycosis, they also infect populations of amphibians throughout much of the world, including parts of Asia, Australia, the United Kingdom, Venezuela, and much of North America. They have been associated with die-offs of many amphibian species, as many types of ranavirus have low species specificity. They have spread throughout the world much like chytridiomycosis—through the international frog trade. Different kinds of ranaviruses produce different symptoms, but most cause tissue necrosis (death), internal bleeding, lethargy, and abnormal wasting of the body, as well as skin sores. These are emerging diseases, the first cases having been discovered in the late 1980s, so little is known about how they are spread from frog to frog, but it is believed that they can remain viable in the environment for long periods of time, even without a host amphibian present.

"Hey—I just spotted a leopard frog!" exclaimed Mike.

"It looks lethargic," said Darin. "It's probably sick."

"Let's catch it and take it back to the field station," said Mike.

The frog died en route to the field station. Once they entered the **necropsy** room of the field station, Mike and Darin donned gloves and started to carefully examine the deceased leopard frog's exterior. It was emaciated and covered in skin ulcers. The tissue of the limbs and tail was necrotic (dead), and the abdomen was edemic (swollen with fluids). The frog was bleeding from the mouth and anus.

"Grosssssssssssss," said Mike. "Let's open this frog up."

Darin took a scalpel and carefully slit the skin of the frog's belly and searched gently for its vital organs. The liver was swollen and friable—it was crumbling apart.

"This doesn't look like chytridiomycosis to me. It's really looking like a ranavirus infection," said Darin. "Let's prepare samples for a virologist and mycologist to confirm what caused this frog to die."

Mike carefully prepared samples of ulcerated skin and tissues of spleen, kidney, liver, and oviduct. Darin bled the frog via **cardiocentesis** (heart puncture). They shipped duplicate samples on ice—one set to a mycologist and the other set to a virologist, Dr. Ann Fischer, who does testing for the U.S. Fish and Wildlife Service.

As suspected by Mike and Darin, the mycologist could not isolate the chytrid fungus from samples nor were antibodies against the fungus found in the blood samples provided.

It took Dr. Fischer several weeks to process the samples. She cultured the samples in the lab, using a fathead minnow epithelial (FHM) cell line. She also fixed the frog's kidney tissues and skin in gluteraldehyde and observed the fixed tissues with the **transmission electron microscope** (TEM). The FHM cells were completely lysed (the cell membrane was broken and the contents spilled out—this is consistent with a common viral life cycle) within 12 days of their inncoluation with the samples. She visualized ranavirus particles in the kidney tissues and in ulcerated skin of the frog. She sent her final report to Mike and Darin.

"You were right, Darin, the frog died of a ranavirus infection," said Mike.

"I knew it! We will have to make another trip to the wetlands to find the source of the virus," said Darin. "We should also find a place to bury the frog's remains."

"Any suggestions?" said Mike.

"I think we should bury the dead frog knee-deep, knee-deep," replied Darin.

Update

Ranavirus mass die-offs in amphibian populations occur in various locations around the world, but most known cases have occurred in North America and the United Kingdom. Ranavirus infections emerged in the United Kingdom in the 1980s. European reports from Denmark, Spain, and the Netherlands note significant declines in the numbers of common frogs, midwife toads, wild alpine and common newts, captive red-tailed knobby newts, wild edible, and water frogs linked to ranavirus infections. Ranavirus infection is considered an emerging disease as listed by the World Organization for Animal Health. Surveillance, control measures, and education of the general public are needed to control its spread. Continued reports about the decline of leopard frogs in the United States may warrant protection under the Endangered Species Act.

Questions to Consider

1. Can humans get the ranavirus in this case study?

2. What do you think the mode of transmission of ranaviruses is and why?

3. Why was a TEM used to visualize ranavirus particles in the frog tissues?

4. What are risk factors for a ranavirus infection?

5. What are the signs and symptoms of a tadpole ranavirus infection and of a frog ranavirus infection?

6. Ranaviruses belong to the *Iridoviradae* family. What are the characteristics of this family of viruses?

7. How stable are ranaviruses in the environment? Can ranaviruses withstand periods of dessication?

8. Explain how the amphibian trade can contribute to the introduction of ranaviruses into new regions.

9. Ranaviral disease is one of two amphibian diseases rated "formidable" and was placed on the Wildlife Diseases List by the World Organization for Animal Health. What does "formidable" mean in this context?

References

Cunningham, A. A. 2007. Emerging Epidemic Diseases of Frogs in Britain are Dependent on the Source of Ranavirus Agent and the Route of Exposure. *Epidemiol Infect* 135: 1200–1212.

Daszak, P., et al. 1999. Emerging Infectious Diseases and Amphibian Populations Declines. *Emerg Infect Dis* 5(6): 735–748.

Duffus, A. L. J. 2010. Major Disease Threats to European Amphibians. *Herpetol J* 20:117–127.

Echaubard, P., et al. 2010. Context-Dependent Effects of Ranaviral Infection on Northern Leopard Frog Life History Traits. *PLoS ONE* 5(10):e13723.

Kik, M., et al. 2011. Ranavirus-Associated Mass Mortality in Wild Amphibians, the Netherlands, 2010: A First Report. *Vet J* doi:10.1016/j.tvjl.2011.08.031.

Relyea, R. A. 2005. The Lethal Impact of Roundup® on Aquatic and Terrestrial Amphibians. *Ecol App* 15:1118–1124.

Relyea, R. A. 2005. The Lethal Impacts of Roundup® and Predatory Stress on Six Species of North American Tadpoles. *Arch Environ Contam Toxicol* 48:351–357.

Relyea, R. A., Schoeppner, N. M., Hoverman, J. T. 2005. Pesticides and Amphibians: The Importance of Community Context. *Ecol App* 15:1125–1134.

Schloegel, L. M., et al. 2009. Magnitude of the US Trade in Amphibians and Presence of *Batrachochytrium dendrobatidis* and Ranavirus Infection in Imported North American Bullfrogs (*Rana catesbeiana*). *Biol Conserv* 142(7):1420–1426.

Ume, Y., et al. 2009. Ranavirus Outbreak in North American Bullfrogs (*Rana catesbeiana*), Japan, 2008. *Emerg Infect Dis* 17(7):1145–1147.

Weldon, C. et al., 2004. Origin of the Amphibian Chytrid Fungus. *Emerg Infect Dis* 10(12):2100–2105.

Internet Resources

Froglife—Information on Ranavirus, http://froglife.org/

Reptile and Amphibian Ecology International: Global Amphibian Population Declines, http://reptilesandamphibians.org/topics/amphibian_declines.html

Chapter 8

Deadly Date Palm Juice

Every culture has its own customs, recreational activities, crafts, and cuisine. In this viral encounter, a local delicacy becomes contaminated with a new and deadly virus that emerged from Malaysia. Finding the reservoir that harbors this pathogen was an epidemiological challenge.

Fred Fischer earned his medical degree in internal medicine, specializing in the diagnosis and treatment of infectious diseases, about 15 years ago. It was always his dream to work for Doctors Without Borders/**Médecins Sans Frontières** (MSF). MSF is a humanitarian organization that brings quality medical care to people whose survival is threatened by violence, neglect, or a catastrophe primarily due to armed conflict, epidemics, malnutrition, exclusion from health care, or natural disasters. The organization sets up refugee camps and field hospitals in 60 countries depending upon the situation and needs.

Dr. Fischer learned about the MSF through one of his colleagues at the Centers for Disease Control and Prevention (CDC). He had been a member of the MSF for over a year and was waiting for a call to action. One day in early January, he was sitting in his office researching about a patient suffering from a mysterious case of **encephalitis** when he got a call from the MSF.

"Can you be in Bangladesh next week?" said Dr. Hossain from MSF headquarters. "Are your vaccinations up-to-date?"

"I sure can!" replied Dr. Fischer enthusiastically. "My vaccinations to travel there are current. I've even had a yellow fever vaccine and Japanese encephalitis vaccine recently."

"Perfect. Remember, although this is one of the coolest months of the year, you'll still need sunscreen."

"I will bring doxycycline (anti-malarial) medication and insect repellent with me too," replied Dr. Fischer.

"The lodging facilities will have insecticide-treated bed nets," said Dr. Hossain. "I'd also suggest you bring some anti-diarrheal over-the-counter medicine."

"What is the situation there?" continued Dr. Fischer.

"People are dying after experiencing high fevers, seizures, and headaches," said Dr. Hossain. "We could really use your infectious disease expertise."

"It sounds like an encephalitis-like illness," pondered Dr. Fischer.

"Yes, you are correct," said Dr. Hossain. "There is an epidemiological investigation under way to determine the risk factors associated with this illness."

A week later, Dr. Fischer was standing in Zia International Airport with a team of doctors and scientists waiting for transportation to Dhaka Hospital where they would be briefed on the situation. Three hours later, they were listening to a presentation given by Dr. Hossain.

"Right now 11 out of 12 people presenting with symptoms have died from encephalitis in the Tangail district. To give you a brief overview on the living situation in that region, the average literacy rate is 30%. About half of the people are involved in agricultural activities, and most of the cases of encephalitis we've seen are farmers. The main agricultural products consist of rice, potato, jute, sugarcane, sesame, linseed, wheat, mustard seed, and pulse, as well as fruit products such as mangos, jackfruit, bananas, litchis, pineapples, and dates. We are quite concerned because there is now an outbreak in a neighboring region with about 40 cases. We think some of these cases may be occurring by human-to-human transmission."

"Have you screened blood samples yet for any of the known infectious agents that cause encephalitis?" interjected Dr. Fischer.

"Yes, we have," continued Dr. Hossain. "It is possible that the farmers are suffering from a viral infection. As you know, there are a number of viruses, including herpesviruses, flaviviruses, and polioviruses that can cause encephalitis. What we're dealing with here appears similar to the Nipah virus, which you may recall caused fatal encephalitis among pig farmers near the town of Ipoh, Malaysia in October 1998. The following February, farmers were dying of a similar encephalitic illness south of the first outbreak in Bukeit Pelandok. At first, we thought the pig farmers and their families in this area had Japanese encephalitis. It was not until March of 1999 that we identified several patients from both areas had been infected with the Nipah virus. "During the 1998–1999 Malaysian outbreaks, we used serology to discover that the Nipah virus produces clinical disease in people, pigs, dogs, and cats, and it can infect horses, sheep, chickens, and bats."

"So, why are we here if we know what is making these people sick?" an American scientist interjected.

"The difference between the Malaysian Nipah virus outbreak and this one is that the sick and dying farmers are not pig farmers," continued Dr. Hossain. "We are beginning to suspect that the common denominator in all of these new cases is that the fresh date palm juice and dates they are consuming are contaminated. This is prime date palm sap collection season, after all. So we have devised a strategy to address the situation from a variety of angles. Our group will be divided into three teams. One team will provide medical care for patients already in our care. Another team will be working throughout the villages, searching for more cases. If more cases are found, the patients must receive supportive care and be put in isolation. The last team will be involved in testing patient serum for antibodies against the Nipah virus and in the field investigating the date palm sap collection process. This team will also be responsible for identifying other potential reservoirs of the Nipah virus causing these outbreaks and alerting local inhabitants to avoid consuming date products until the outbreak has been controlled."

Dr. Fischer was assigned to the third team, whose task it was to identify potential Nipah virus reservoirs. He was ecstatic to finally get out in the field. The next day he teamed with an Indian scientist, Dr. Nahar, and a local translator to investigate the date palm sap collection process. Very little information is available about the harvesting of sap from palm trees. They began in-depth interviews with the date palm sap collectors, also called *gachhis,* all working in one area that had 12 date palm trees.

They observed a bare-footed *gachhi* climbing up a date palm tree. A rope was wrapped around his waist, fastening his legs to the tree to aid in climbing and to avoid accidents. He was carrying a basket tied to the rope, containing a knife, a sickle, and hollow bamboo sticks. A pitcher was also hung on the rope. The *gachhi* cleaned the tree's surface with the sickle, then used his knife to create a V-shaped cut into the tree. He inserted a thin hollow bamboo stick into the V-shaped cut. An earthen pot was placed under the outer end of the bamboo stick, collecting the sap dripping from the tree. Then the *gachhi* secured netting over the shaved part of the tree and the collection jar (**Figure 4**).

Drs. Fischer and Nahar were fascinated by the agility of the **gachhi.** He made the collection process look effortless. After observing the date palm trees, it appeared that the sap was collected from the opposite side of the previous year's cut. The *gachhi* quickly climbed down the date palm tree. The translator approached him and explained the recent outbreaks of encephalitis that may be tied to the process of collecting the sap and asked if he would be willing help them determine the exact cause. The *gachhi* was willing to answer their questions.

"What's the cloth netting for?" asked Dr. Fischer.

Figure 4. Gachhi collecting date palm juice. (Drawn by Brian Ledwell.)

The translator questioned the *gachhi*, and he responded with intensity and strong gesturing. He was very animated about something.

"The *gachhi* says that the bees and hornets get into the sap, which reduces its value. The cloth or bamboo netting is used to keep the bugs out. He says that birds, dogs, foxes, and bats drink the sap occasionally and spoil it with urine and feces. The sap becomes cloudy and smelly, and they can't sell it for a good price. This time of year, sap collection is their livelihood. They make molasses from the sap, as well as using it as a sweetener in traditional cakes and desserts. And often the sap is eaten raw," continued the translator. "Anyway, as I was saying, the birds and animals can spoil the sap and can also break the collection pitchers. So he is trying hard to prevent the pests from spoiling the date palm juice so that it can be sold at a higher price. Besides netting, *gachhis* also use thorns and branches to keep the animals out of the sap, or they narrow the mouths of the pots. Some *gachhis* even add lime to the collection pots and the cuts in the trees to help keep the sap clear."

"It looks like there could be several pests that contaminate the sap with Nipah virus and other microbes when they feed on the sap," said Dr. Nahar. "I wonder if the *gachhis* also taste the sap themselves; that could be how they are getting sick," verified Dr. Nahar. "Perhaps this gentleman will allow us to take some sap samples? Then tomorrow we can team up with the ecologists to trap target animals for testing," continued Dr. Nahar.

For the next few days, with the help of a local translator, the ecologists, along with Drs. Fischer and Nahar trapped bats and other animals who frequented the sap pots. Since Nipah virus is classified as a biosafety level-4 (BSL-4) pathogen, the entire team wore protective clothing such as impermeable gloves, masks, goggles, and boots during their fieldwork. They collected blood, urine, and saliva samples from the animals and swabbed harvesting tools used for sap collection and various batches of sap. Samples were sent to a laboratory for testing.

Two weeks later, Dr. Fischer was back in his practice, chatting on the phone with Dr. Nahar. "All of the fruit bat blood samples were seropositive for Nipah virus. The virus was isolated by cell culture and confirmed with polymerase chain reaction (PCR) assays," said Dr. Nahar.

"We really need to create a vaccine against Nipah virus," said Dr. Fischer. "Right now all we can do is intensive supportive care to treat the symptoms."

"The mortality rate can be as high as 92%!" said Dr. Nahar.

"The outbreaks appear to be seasonal, coinciding with sap collection, as we had proposed," added Dr. Fischer.

"The *gachhis* in this Nipah-belt will continue to be exposed to the bats carrying the virus during the harvesting of the date palm sap," said Dr. Nahar. "We need to work with the *gachhis* in developing effective means of keeping the bats away from the date palm trees and themselves."

"It will be important for us to work with the locals to address this problem, especially until a vaccine can be developed," replied Dr. Fischer.

"Yes, and we'll have to learn and understand the local beliefs and practices of the *gachhis*," said Dr. Nahar. "Sap collection is their livelihood, so instead of trying to make them change their ways, we'll have to help them devise ways of collecting the sap more safely."

"I would love to travel to Bangladesh during the winter season to assist the locals with sap collection methods!" said Dr. Fischer. "If I work at my practice every day here in the United States, I'll go batty!"

Update

Since Nipah virus was first discovered in 1999 during an outbreak among pig farmers in Malaysia, there have been 14 more outbreaks, all in South Asia. Efforts are underway to decrease bat access to date palm sap. Freshly collected date palm juice should also be boiled, and the dates thoroughly washed and peeled before they are eaten.

Questions to Consider

1. Define encephalitis. List other viruses that cause encephalitis.

2. Nipah virus is closely related to Hendra virus. Research Hendra virus. What illness does it cause? What hosts does it infect? Where do Hendra virus infections occur?

3. Research on Nipah virus is performed in a BSL-4 laboratory. What is a BSL-4 laboratory?

4. Besides Nipah virus, what other viruses can fruit bats carry and transmit to humans?

5. What changes in recent ecological changes have occurred that have resulted in the fruit bats increasingly coming into contact with humans and domesticated animals?

6. Hendra and Nipah viruses are closely related paramyxoviruses. What are the molecular characteristics of paramyxoviruses?

References

Bossart, K. N., et al. 2007. Targeted Strategies for Henipavirus Therapeutics. *Open Virol* 1:14–25.

Chong, H. T., et al. 2008. Differences in Epidemiologic and Clinical Features of Nipah Virus Encephalitis Between the Malaysian and Bangladesh Outbreaks. *Neurol Asia* 13:23–26.

Johara, M. Y., et al. 2001. Nipah Virus Infection in Bats (Order Chiroptera) in Peninsular Malaysia. *Emerg Infects Dis* 7(3):439–441.

Khan, M. S., et al. 2010. Use of Infrared Camera to Understand Bats' Access to Date Palm Sap: Implications for Preventing Nipah Virus Transmission. *Ecohealth* 4(4): 517–525.

Lo, M. K., et al. 2012. Characterization of Nipah Virus from Outbreaks in Bangladesh, 2008–2010. *Emer Infect Dis* 18:248–255.

Luby, S. P., et al. 2006. Foodborne Transmission of Nipah Virus, Bangladesh. *Emer Infect Dis* 12(12):1888–1894.

Nahar, N., et al. 2010. Date Palm Sap Collection: Exploring Opportunities to Prevent Nipah Transmission. *Ecohealth* 7(2):196–203.

ProMED-mail. Undiagnosed Deaths, Encephalitis—Bangladesh (Kushtia):Nipah, ProMED-mail 2007;25 Nov:20071125.3816.

ProMED-mail.Nipah Encephalitis, Human—Bangladesh: Nipah ProMED-mail 2011;8 March:20110308.0756.

Rahman, M. A., et al. 2012. Date Palm Sap Linked to Nipah Virus Outbreak in Bangladesh, 2008. *Vect Borne Zoonotic Dis* 12(1):65–72.

Stone, A. 2011. Breaking the Chain in Bangladesh. *Science* 331:128–1131.

Vector-Borne Disease Control Section, et al. 1999. Update: Outbreak of Nipah Virus—Malaysia and Singapore, 1999. *MMWR* 48(16):335–337.

Internet Resources

Doctors Without Borders//Médecins Sans Frontières, http://doctorswithoutborders.org

About Doctors Without Borders on YouTube, http://www.youtube.com/watch?v=73zMcdGfXGE&feature=related

Hendra and Nipah Viruses, CDC Special Pathogens Branch, http://www.cdc.gov/ncidod/dvrd/spb/mnpages/dispages/nipah.htm

Traveler's Health CDC home page, http://wwwn.cdc.gov/travel/destinations/list.aspx

WHO Nipah fact sheet, http://www.who.int/mediacentre/factsheets/fs262/en/

Chapter 9

Snotty Horses

Roman chariot races are one of the earliest examples of horse racing. Today, Australia is one of the top three thoroughbred racing nations of the world. It is a billion dollar industry. On August 25, 2007, only a week before the start of a multi-million-dollar racing and thoroughbred-breeding season, a viral outbreak spread through the horses. The outbreak halted horse importation and forced the cancelation of a 3-day qualifying event for the 2008 Summer Olympics scheduled in Sydney.

Joe Keech was the main veterinarian at the Eastern Creek Quarantine Centre in Sydney, Australia. At the moment, he was puzzled. Secretariat, a racing stallion, had a cough. The horse's eyes were tearing, and his nose was watering and exuding mucus. He had a soaring fever of 105° F (healthy horses have a temperature range of 99.5° F to 100.5° F. The horse was lethargic. He wouldn't eat and his breathing was labored. Keech thought it was pretty likely that the horse had equine influenza, but whatever Secretariat had, its onset was sudden. Australia had the best quarantine system in the world, and the horse hadn't had any symptoms when he'd come into the country only days before.

Joe drew some blood and performed a nasal swab for testing. Then he called the horse's owner. "Secretariat is sick," Joe said with trepidation.

Secretariat's owner replied, "What's wrong with him?"

Joe hesitated, "I'm not sure but we need to isolate him. I am so sorry, sir."

"This horse is in training for a race next week!" yelled the owner. "This timing will cost me lots of money!"

"We can't exercise Secretariat, sir," said Joe. "He can't be transported."

Joe was timid in dealing with the horse's owner. The rest of the conversation was very tense. The owner was upset. Joe was upset. Joe decided to call his friend Bill Roberts, a vet at Sydney's Centennial Parklands Equestrian Centre. "All the signs and symptoms point to equine influenza," said Joe, after describing Secretariat's symptoms.

"What a coincidence you should call!" said Bill. "I've got some coughing stallions in my care as well."

"I've been researching any recent RFIs (Request for Information) in **ProMED** and found one on an outbreak that is going on in Japan right now," said Joe. "It appears to be equine influenza, but it's not confirmed yet."

"Well, there are horses imported from Japan, the United Kingdom, Ireland, and the United States that are quarantined in Sydney right now," added Bill.

"Our Australian horses are all susceptible because we don't have a mandatory vaccination program," replied Joe. "Equine influenza is highly contagious!"

A few days passed. Laboratory tests confirmed that the sick horses in Joe's and Bill's care had equine influenza A, sub-type H3. Bill and Joe were involved in an emergency animal disease response, vaccinating and checking horses around New South Wales. The Australian Animal Health Laboratory carried out testing. Horse importation was banned, and horse owners were told to abide by this ban by keeping their horses on their properties. Local news reports focused on the cancellation of race meetings. The public was asked to report all horses with a fever and respiratory signs to the Local Disease Control Center in Queensland.

"Did you hear that at least 20 individuals were intercepted while trying to illegally transport horses since the equine ban was initiated?" asked Joe during a phone conversation with Bill.

"Even though they face a maximum penalty of $150,000 or 2 years in jail! It's really hard to get owners and trainers to comply."

"The standstill is very strict," added Joe. "No horses of any type—pets, riding club horses, show animals, donkeys, mules, or zebras can be moved."

"Even cattle mustering, which is carried out on horseback, has been halted in Queensland because of this lockdown," replied Bill.

"The outbreak is forcing cancelation of all horse-related events," Joe concurred.

"The racing industry is crippled right now!" exclaimed Bill. "Trainers aren't permitted to have any contact with the owners and the owners, in turn, won't be permitted within 20 meters of their horses."

"Authorities still don't know how the virus got past the quarantine procedure," said Joe. "My guess is that the virus was transmitted on clothing or equipment of the trainers or other people who had contact with imported sick horses."

"I think there was a contaminated horse trailer," speculated Bill. "Too bad the horses weren't vaccinated properly."

"Vaccines don't give absolute protection," said Joe. "The viruses mutate to the point in which the horse's immune system no longer recognizes the mutated virus."

"Then the vaccines need to be updated to keep up with the changing virus," said Bill.

"It's been taking my sick horses about two weeks to recover," said Joe.

"Some of the horses are getting a secondary bacterial infection. Their snot is yellow or green," replied Bill. "I need to get back to treating the horses under my watch."

"Yeah, time to quit horsing around," said Joe with a smile. "We both need to get back to the stables."

Update

Over 50,000 horses contracted equine influenza during the Australian outbreak. Some fatalities were reported, but the actual number is unknown. It appeared to be very low. Commissioners criticized Australia's quarantine system as being "inefficient, underfunded, and lacking diligence." The Australian government pursued a goal of eradication and return to an equine-influenza-free status. The outbreak and subsequent event cancellations cost the Australian racing industry at least 40 million dollars. Infectious respiratory disease continues to constitute one of the major health and economic threats to the horse industry.

Questions to Consider

1. What is the best prevention against the introduction of equine influenza?

2. Do all countries mandate quarantine and vaccination systems designed to stop viral diseases? Explain your answer.

3. How could this disease start in a quarantine station and outside of it?

4. How is equine influenza spread?

5. Equine influenza is rarely fatal in horses. Which horses would be more likely to experience complications or die of this respiratory disease?

6. Can humans and other animals besides horses get equine influenza?

7. How does equine influenza A differ from human influenza A?

8. Equine influenza is an enzootic disease. Define enzootic.

References

Daly, J. M., et al. 2004. Current Perspectives on Control of Equine Influenza. *Vet Res* 35(4):422–423.

Elton, D., Bryant, N. 2011. Facing the Threat of Equine Influenza. *Equine Vet J* 43(3):250–258.

Gildea, S., et al. 2011. The Molecular Epidemiology of Equine Influenza in Ireland from 2007–2010 and Its International Significance. *Equine Vet J*. doi. 10.11.1111/2042-33062011.00472.

Paillot, R., et al. 2006. Vaccination Against Equine Influenza: Quid Novi? *Vaccine* 24(19):4047–4061.

ProMED-mail. Equine Influenza—Australia EX Japan. ProMED-mail 2007;25 Aug:20070825.2798.

ProMED-mail. Equine Influenza—Australia (New South Wales, Queensland). ProMED-mail 2007;20071017.3394.

Rezendes, A. 2007. Australia Battles Equine Influenza. *J Amer Vet Med Assoc* 231(8):1189.

Internet Resources

DEFRA/AHT/BEVA Equine Quarterly Disease Surveillance Reports, http://www.aht.org.uk/cms-display/disease_surveillance.html

American Association of Equine Practitioners, http://www.aaep.org/

Worms and Germs blog, http://www.wormsandgermsblog.com/

Chapter 10

A New Route of HIV Transmission?

In this viral encounter, HIV transmission is linked to a customary practice used by mothers in various parts of the world, including the United States. Two children in Florida and a child in Tennessee were infected with HIV this way.

Dr. Ann Ferran was a virologist and associate professor in the Biology Department at Sun College. She had been lecturing to her students about HIV all week and was excited to discuss current research pertaining to a new mode or risk for HIV transmission. She began lecturing with a review of the previous lecture.

"First, let's list the **modes of transmission** of **HIV** on the board," she asked the class. A few hands were raised.

"Sex," said Justin Brown. Dr. Ferran began writing on the board as students contributed answers. "Even oral sex and French kissing if an HIV-positive person has bleeding gums and the partner has bleeding gums or sores in their mouth."

"Blood transfusions," said Sarah Novack.

"That's pretty rare in the United States," added Dr. Ferran. "Remember that the blood supplies, including clotting factors, are screened for HIV before donation."

"Sharing needles or syringes for drug injection," said Jennifer Cunningham.

"What about that dentist in Florida?" said John Schultz. "Didn't he infect a few patients with HIV?"

"Those patients had tooth extractions," said Kari Kromm. "The dentist was HIV positive."

"I thought nobody knows for sure how those patients got HIV?" said another student.

"That's right," replied Dr. Ferran. "What was learned during the Centers for Disease Control and Prevention's (CDC's) investigation?"

"All three patients had a similar strain of HIV in their bodies," chimed in a student in the back of the room.

"That the staff members, including the dentist, practiced barrier precautions while treating patients," said Sarah.

"They washed their gloves instead of changing them between patients," said Tom Keller. "And they wore masks but didn't change them very often."

"The autoclave used to sterilize their instruments was working just fine," added another student.

"So...suggest a *likely* mode of transmission," said Dr. Ferran.

"The dentist used needles to administer novacaine anesthetic," said Jennifer. "I'm betting the dentist had an accidental needle stick injury."

"Can you expand upon that, Jennifer?" said Dr. Ferran.

Jennifer responded, "The dentist's blood during a needle stick injury might have gotten into the patient during a procedure like a tooth extraction."

"Good point, but the truth is, we can't be certain exactly how HIV was transmitted in these instances," said Dr. Ferran. "This is a great discussion for later, but let's keep working on our list of HIV modes of transmission on the board."

The class was quiet. Dr. Ferran prodded the class along. "How do babies born to HIV-infected mothers become infected?" she asked.

"During the birthing process," said Robert Ledwell. "Or through breastfeeding after birth."

"Excellent!" said Dr. Ferran. "Now I want to discuss three very interesting cases since 1993 in which children became infected with HIV through a practice called **pre-mastication**."

"What is pre-mastication?" asked a student in the front of the room.

"It's a practice when mothers or caregivers chew an infant's food," answered Dr. Ferran. "It makes the food more palatable for children." While the students looked at each other with expressions of disgust, she continued, "In developing countries, dental care is lacking and there are not commercially available baby foods and blenders. Parents and caregivers may need to soften the foods, so they chew the foods for the children. It's taken researchers several years to connect these cases and to report their findings."

Dr. Ferran continued to outline the cases for her students. "All three children tested negative for HIV after birth." She displayed a PowerPoint slide that compared all three cases (**Figure 5**).

Case 1 (Memphis)	Case 2 (Miami)	Case 3 (Miami)
African American female child HIV– at birth	African American male child HIV– at 20 and 21 months	African American male child no HIV status at birth
Father HIV–	Father HIV–	Father HIV–
Mother HIV-1+, on HAART therapy during pregnancy and postpartum; child delivered by caesarean; on liquid AZT for first 6 weeks of life; denied breast feeding	Mother HIV-1+	Mother HIV–
Child HIV-1 + 9 months old (2004)	Child HIV-1 + 39 months old (1995)	Child HIV-1 + 15 months old (1993)
Mother began giving child pre-masticated food at 120 days of life	Mother offered child pre-masticated food but could not provide time frame to physicians	Mother tested HIV– three times at the time the child tested HIV+
		Child spent time with great aunt who was HIV+. She was in a 12-year relationship with HIV+ man who had a history of intravenous drug use. The child was 9–14 months when she acted as a caregiver; the great aunt gave the child pre-masticated food.

Figure 5. Dr. Ferran's PowerPoint slide outlining three cases.

Dr. Ferran carefully went through each case. Then she asked the class a challenging question, "Besides pre-mastication, what other modes of transmission could be used to explain the HIV+ status of these children?"

The class was silent. Then Kari Kromm said, "Did any of the children have a blood transfusion?"

"No, but that is a good suggestion," said Dr. Ferran.

"How about an organ transplant?" said another student.

"No," said Dr. Ferran.

"Did any of the children receive any medication intravenously?" asked Jennifer Cunningham.

"There is no record of it," said Dr. Ferran.

"How about sexual abuse?" asked a student very quietly in the back of the room.

"That's a possibility but again, it was ruled out for lack of evidence," replied Dr. Ferran.

"Maybe there were needle sticks with blood from the druggie boyfriend in case 3?" said a student in the front of the room.

"Good point, but the great aunt denied seeing any needles around the house and that the child was ever stuck by one," said Dr. Ferran.

The class was quiet. "Can anyone think of any other possibilities?" said Dr. Ferran. Students in the class shook their heads no. Some surfed the Internet for ideas using their laptops, tablets, iPods, and smart phones. Dr. Ferran paused and then said, "Neither could the doctors investigating these cases."

"I still don't understand how pre-mastication could cause the child to become HIV infected," asked Kari Kromm. "How was the virus transmitted through the food?"

"I bet the mothers or the great aunt's gums were bleeding or they had sores in their mouths when they chewed the food for the children," said Jennifer Cunningham.

"Good point. The mother in the first case had gums that bled spontaneously," said Dr. Ferran. "There is no oral history for the mother in the second case. The great aunt's gums in the third case bled very often, and she probably had open sores in her mouth. When the child was 14 months old, the great aunt died of pneumonia," continued Dr. Ferran. "The blood and saliva from the HIV-1+ mothers and the aunt caregiver contained the HIV-1 virus. The children were given pre-masticated food at a time when they were teething, when the child's gums may have been inflamed or bleeding, and they may even have had mouth sores, allowing the virus to enter via the pre-masticated food that contains HIV in the blood or saliva of the HIV-infected mothers and great aunt. Therefore, the investigators concluded that pre-mastication put these children at risk for HIV infection."

"What a fascinating cases!" said John Schultz. "Have kids got infected with other pathogens through the pre-mastication of food?"

"That's a great question!" said Dr. Ferran. You could hear students typing on their laptops, tablets, and smartphones, surfing the Internet again.

"I want everyone to research this topic, and we will discuss it in the next lecture," said Dr. Ferran. "We've only got a few minutes to quickly wrap up today's lecture."

Update

The results of the pre-mastication study were published in 2011. The practice of pre-mastication continues to occur in the United States, where at least 14% of caregivers admit to pre-masticating food for children. It is important for HIV-infected caregivers to be aware that they should not pre-masticate for any child.

Questions to Consider

1. What countries/cultures practice pre-mastication?

2. What body fluid usually contains more HIV virus—blood, sweat, tears, or saliva? Is HIV more likely to be transmitted by blood, sweat, tears, or saliva? Explain your answer.

3. How stable is HIV in the environment? What is the best way to disinfect households that may contain the HIV virus?

4. What other pathogen(s) can be transmitted through the practice of pre-mastication.

5. List the different tests for HIV-1. Discuss what these tests are detecting and what they entail.

6. What is AIDS? What are the signs and symptoms of AIDS?

7. List all possible routes of HIV transmission.

8. Could hepatitis B virus be transmitted by pre-mastication? Research this in the literature using PubMED or another literature search engine.

References

Brown, L. M. 2000. *Helicobacter pylori*: Epidemiology and Routes of Transmission. *Epidemiol Rev* 22:283–297.

Centers for Disease Control and Prevention. 1991. Epidemiologic Notes and Reports Update: Transmission of HIV Infection During Invasive Dental Procedure—Florida. *MMWR* 40(2):21–27, 33.

Centers for Disease Control and Prevention. 2011. Premastication of Food by Caregivers of HIV-Exposed Children—Nine U.S. Sites, 2009–2010. *MMWR* 60(9):273–275.

Clemens, J., et al. 1996. Sociodemographic, Hygienic and Nutritional Correlates of *Helicobacter pylori* Infection of Young Bangladeshi Children. *Pediatr Infect Dis J* 15: 1113–1118.

Gaur, A.H., et al. 2009. Practice of Feeding Pre-masticated Food to Infants: A Potential Risk Factor for HIV Transmission. *Pediatrics* 124(2):658–666.

Hafeez, S., et al. 2011. Infant Feeding Practice of Premastication: An Anonymous Survey among Human Immunodeficiency Virus-Infected Mothers. *Arch Pediatr Adolesc Med* 165(1):92–93.

Hillis, D.M., Huelsenbeck, J. P. 1994. Support for Dental HIV Transmission. *Nature* 369:24–25.

Ivy, W. 3rd, et al. 2012. Premastication as a Route of Pediatric HIV Transmission: Case-Control and Cross-Sectional Investigations: Pediatric HIV Risk Via Premastication. *J Acquir Immune Defic Syndr* 59(2):207–212.

Lindkvist, P., et al. 1998. Risk Factors for Infections With *Helicobacter pylori*—A Study of Children in Rural Ethiopia. *Scand J Infect Dis* 30:371–376.

Maritz, E. R., et al. 2011. Premasticating Food for Weaning African Infants: A Possible Vehicle for Transmission of HIV. *Pediatrics* 128(3):e579–e590.

Megraud, F. 1995. Transmission of *Helicobacter pylori*: Faecal-Oral Versus Oral-Oral Route. *Alimen Pharmacol Therap* 9(Suppl. 2):85–91.

Ou, C.-Y., et al. 1992. Molecular Epidemiology of HIV Transmission in a Dental Practice. *Science* 256:1165–1171.

Internet Resources

CDC HIV/AIDS, http://www.cdc.gov/hiv/

WHO HIV, http://www.who.int/hiv/en/

PubMED home page, http://www.ncbi.nlm.nih.gov/pubmed/

Chapter 11

Zombie Fish

Media reports about this deadly Ebola-like virus that kills at least 50 different marine and freshwater fish species made the headlines of U.S. newspapers from 2006–2007. It continues to spread affecting several water systems in the United States and Canada.

"Hey guys, what are you doing?" asked Tanner Mesko in a low voice. Tanner watched two men in their twenties drive up to a Lake Winnebago boat launch. The launch ramp was located in Menominee Park in Oshkosh, Wisconsin. The boat was wet inside, and weeds were hanging from it. Tanner noticed the license plate on the boat trailer was from Michigan.

"We're going fishing, what do you think we're doing?" snarled one of the men sarcastically.

Tanner approached the two young men. "Sorry to startle you two," he said calmly. "Where are you guys from?"

"Michigan. The fishing sucks there," said the guy with a Detroit Tigers baseball cap. "We've been launching into whatever lake we find."

"Yeah, we've been doing this for 4 weeks. Might as well, since we're not working," said his bearded fishing partner." He paused, "Like, we are unemployed."

"We used to work at the G.M. factory in Detroit."

Tanner knew the car manufacturers were struggling these days. The economy was in a slump. Both of them looked unkempt and scruffy, like they had been camping and partying for a few weeks.

"We just came from Rush Lake," said the bearded guy.

Tanner knew Rush Lake. It was about a 30-minute drive away and not part of the Lake Winnebago system. As soon as he learned that the guys were lake-hopping, he was very concerned. Tanner was an avid fisherman and hunter and was well educated about fish diseases. His wife was a microbiologist who was always informing him about infectious diseases of fish and other wildlife. He knew how easily diseases could spread from one water system to another.

In attempt to get more information out of the men, Tanner introduced himself. "My name is Tanner, I live here in Oshkosh." After a moment, he added on, "Have you two caught any sick fish?"

"I'm Alex," said the guy with the Detroit Tigers hat.

"And I'm Tom," said the bearded guy. "You know, Alex hooked a bluegill the other day that had these bulging eyes. It sure didn't look healthy."

"Yeah, It looked like a zombie fish! There was blood coming out of its eyes and it was all bloated," added Alex.

"We kept it on the boat for a while but it started to stink, so we threw it into some other lake along the way," added Tom.

"I saw your Michigan plates," replied Tanner. "I figured you would know about VHS since it's in the Great Lakes."

"VHS?" asked Alex. "Is that some kind of videotape?"

Tanner sighed. He could clearly see that Alex and Tom had no idea about fish diseases and were reckless and irresponsible regarding their fishing habits.

"VHS stands for viral hemorrhagic septicemia," said Tanner. "It's a fish disease caused by a virus. It's causing die-offs in the Great Lakes region and has been spreading fast."

"Tom, your eyes are blood shot," Alex said with alarm. "I told you not to keep that bug-eyed bluegill!"

"Humans can't get it," interrupted Tanner. "But you can spread it from lake to lake by not cleaning your boats after you fish."

"Excuse me, are you accusing us of not cleaning the boat?" asked Tom.

"Well, I see Eurasian milfoil and zebra mussels hanging from the propeller of the boat," replied Tanner. "They are invasive species that can also be spread if you don't clean and disinfect your boat properly. Your boat also isn't drained, and the fishing equipment doesn't look disinfected," continued Tanner.

Tom whispers to Alex, "This guy is annoying, let's get out of here." The two of them got back into the truck that still had the trailer with the boat attached and sped off. Tanner could hear the tires screeching as they sped out of the parking lot.

Tanner was upset. He decided to contact people at the Wisconsin Department of Natural Resources (DNR). He spoke with Bill Smith, a member of the VHS Response Team who was involved in fish-health compliance and tracking fish kills. Bill assured him that they would be watching for Tom and Alex and would intervene to address their reckless behavior.

"They have violated at least one the VHS emergency rules," said Bill. "Failing to drain all water from the boat, trailer, or fishing equipment can cost around $329 in fines. They could lose their fishing licenses as well. All of the statutes are posted on the DNR's web page."

Three weeks passed. Tanner took a weekend fishing trip with his two sons to Lake Plentiful located in central Wisconsin. It was a pristine lake, not part of any river or other lake system in the state. The lake was known as a great fishing spot for its plentiful stocks northern pike, bluegill, crappies, walleye, and bass. Rumor had it that the DNR was even going to stock this lake with rainbow trout, including trophy specimens.

It was a calm, cool, day—perfect for catching bluegills. Tanner and his sons launched the boat and motored across the lake to the shallower end where they had had great luck with catching bluegills in the past. Once they reached their destination, they couldn't believe what they were seeing: a huge expanse of dead and dying bluegill and crappies. There must have been at least 1000 dead or sick fish. The fish that were alive looked pale and anemic, they were swimming abnormally, and their eyes and the bases of their fins were hemorrhaging. Their bellies were distended. Normal die-offs do occur in the spring, but this was not a normal die-off; it looked like fish suffering from disease. This should be a VHS-free lake!

Fifty yards away, they observed a DNR boat. It looked like two individuals aboard were netting the sick fish. Tanner decided to motor over to it and find out what was happening.

"What's wrong with the fish?" Tanner asked the DNR agents.

"We're not sure. That's why we're collecting fish and water samples for testing," responded one of the men. "Just when you think this lake is protected, a threat of disease sets in."

"What fish diseases will you be testing for?" asked Tanner.

"VHS and IPN," replied one of the DNR officials.

Tanner's son asked, "What is IPN?"

"Infectious pancreatic necrosis," replied the DNR official. "It's a fish disease caused by a virus."

"It can infect rainbow trout, and we had plans to stock this lake, so we need to find out quickly if the lake is infected." added the other DNR official. "IPN was just found in a local hatchery."

"What kinds of lab tests will you be doing?" asked Tanner.

"We will try to isolate virus from these fish and water samples and grow them in BF-2 (bluegill fry), RTG-2 (rainbow trout gonad), or EPC (epithelioma papulosum cyprini) or FHM (fathead minnow) cells," replied one of the officials. "We can also confirm virus identity by **virus neutralization**, immunofluorescence (IFA), an **enzyme-linked immunosorbent assay** (ELISA), or a polymerase chain reaction (PCR)-based test."

"Whoa, sounds complicated," said Tanner. "My wife is a microbiologist; I'll ask her to explain all of this science stuff."

One of the DNR officials said, "I should introduce myself. I am Bill Smith and this is my partner Jack Miller."

"Bill Smith? I spoke with you on the phone a few weeks ago about two young men from Michigan lake-hopping and not disinfecting their boat," said Tanner.

"Oh yeah, we caught up to those two," said Bill.

"We cited them for a number of statute violations," added Jack Miller. "They didn't drain and disinfect their boat or disinfect their fishing gear. Plus, they were moving live fish and fish eggs from one lake to another."

"And they were cited for the unlawful possession and use of dead fish," said Bill. "They didn't even have fishing licenses!"

"We were able to document their behavior on a camcorder as they were messing around on this lake a few weeks ago," said Jack.

"The incubation period for VHS virus in freshwater is 1 to 3 weeks," said Bill. "It's looking like they or someone like them might be the culprits in spreading VHS to this lake."

"The virus is shed into the water through urine and reproductive fluids, which is why we are taking water samples for testing," added Jack. "The virus can survive in the water for at least two weeks."

"I don't remember hearing anything about VHS until the last few years," said Tanner.

"VHS became a subject of media reports in 2006," said Bill. "Genetic tests show that the strain of VHS found in the Great Lakes probably originated in Atlantic Ocean fish near New Brunswick, Canada."

"It's likely the VHS strain started infecting fish in the St. Lawrence River, which is a shipping route that leads to the Great Lakes," continued Jack. "Now this fish pathogen is found in the Great Lakes, including Lake Michigan, and made its way to the Lake Winnebago System in 2007."

"We are doing testing to determine if Lake Superior and the Mississippi River System have VHS-infected waters," added Jack.

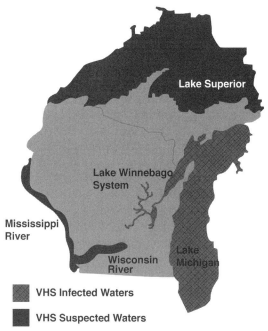

Figure 6. VHS-infected and VHS-suspected waters in Wisconsin. (Adapted from WDNR. Drawn by Brian Ledwell.)

"Well, I guess we'll have to disinfect our boat and try to fish on another day," said Tanner. "There's nothing fishy going on here."

Update

*VHS is a deadly infectious fish disease that has continued to spread among freshwater fish in the Great Lakes region since 2003. The virus found in Wisconsin is a new genetic strain that is most closely related to the virus found in the Pacific Northwest, rather than a strain found in Europe. **Figure 6** illustrates VHS-infected waters and VHS-suspected waters in the Wisconsin vicinity.*

Questions to Consider

1. How does the VHS virus spread from fish to fish?

2. Can some fish become carriers of VHS? Why or why not?

3. List the species of fresh and marine fish that are susceptible to VHS.

4. List ways that the spread of VHS can be prevented.

5. Are VHS-infected fish safe to eat? Explain your answer.

6. Compare and contrast the signs and symptoms of infectious pancreatic necrosis versus viral hemorrhagic septicemia.

References

Ammayappan, A., Vakharia, V. N. 2009. Molecular Characterization of the Great Lakes Viral Hemorrhagic Septicemia Virus (VHSV) Isolate from USA. *Virol J* 6:171.

Einer-Jensen, K., et al. 2004. Evolution of the Fish Rhabdovirus Viral Haemorrhagic Septicemia Virus. *J Gen Virol* 85:1167–1179.

Faisal, M., Winters, A. D. 2011. Detection of Viral Hemorrhagic Septicemia Virus (VHSV) from *Diporeia* sp. (*pontoporelidae, amphipoda*) in the Laurentian Great Lakes, USA. *Parasit Vectors* 4:2.

Gagne, N., et al. 2007. Isolation of the Viral Haemorrhagic Septicemia Virus from Mummichog, Stickleback, Striped Bass and Brown Trout in Eastern Canada. *J Fish Dis* 30(4):213–223.

Hope, K. M., et al. 2010. Comparison of Quantitative RT-PCR with Cell Culture to Detect Viral Hemorrhagic Septicemia Virus (VHSV) IVb Infections in the Great Lakes. *J Aquat Anim Health* 22(1):50–61.

ProMED-mail. Infectious Pancreatic Necrosis, Trout—USA (Minnesota Ex Wisconsin) 2009; 10 Feb:20090210.0601.

ProMED-mail. Viral Hemorrhagic Septicemia, Fish—USA (Michigan, Wisconsin) 2007; 19 May:20070159.1595.

Internet Resources

Viral Hemorrhagic Septicemia VHS—WDNR, http://dnr.wi.gov/fish/vhs/vhs_widistribution.html

Fish Diseases: Viral Hemorrhagic Septicemia: Minnesota DNR, http://www.dnr.state.mn.us/fish_diseases/vhs.html

Chapter 12

Mysterious Bleeding

During September and October of 2008, an aggressive and lethal new virus caused a mysterious outbreak in Zambia. This case is an example of how fast a new virus can be identified—a process that used to take weeks, months, or even longer.

A female, safari travel agent living in Lusaka, Zambia suddenly felt very ill. She had flu-like symptoms—high fever, chills, vomiting, and diarrhea. It took all the energy she had to get herself to a hospital.

She was placed in an intensive care unit, and doctors and nurses provided supportive care, but still her conditioned worsened. She developed a rash, her liver was starting to fail, and she was suffering from periodic convulsions.

"She's bleeding from her gums and from intravenous injection sites," said the Zambian doctor to the nurse in the room. "Have we received any test results from the medical technology lab? That would help us narrow down what she has. Perhaps it could be tick-bite fever?"

The medical technologist tested for disease agents in the sample provided. He quickly ruled out *Rickettsia africae*, the bacterium that causes *tick-bite fever*. What else could it be? "She might have a viral hemorrhagic fever," said the Zambian doctor, studying the patient's charts. "She keeps getting worse, and we can't get her bleeding under control. We can't care for her effectively here. Let's evacuate her to a hospital in Standton, down near Johannesburg. They are better equipped to deal with this critically ill patient and to determine the cause of her illness."

The hospital was buzzing with phone calls. "We need a paramedic to transport her to the aircraft."

She was rushed on a gurney into a helicopter with a male paramedic attending to her during the flight to Johannesburg. The paramedic struggled to control the bleeding.

After their arrival at the hospital in Standton, healthcare workers wearing masks, gloves, and gowns—the **barrier technique** for biohazard protection—surrounded her. Her condition continued to decline quickly. She became comatose, went into organ failure, and died a day after her arrival.

Nine days later, the paramedic who had taken care of her in the helicopter, developed nonspecific flu-like symptoms: headache, fever, and muscle pain. Then he progressed to similar symptoms as the travel agent: high fever, chills, vomiting, and bleeding from his gums. The Zambian doctor sent a blood sample to the medical technologist and called him on the phone. "I am sending you some blood samples from a patient we suspect may have a viral hemorrhagic fever," said the doctor. "It could be the same disease that affected the patient whom I thought had *tick-bite fever*."

"There's never been a case of Ebola hemorrhagic fever in Zambia if that's what you're thinking," said the medical technologist. "Besides, this lab isn't equipped for diagnosing viral hemorrhagic fevers such as Ebola virus infections,"

The doctor replied, "The patient's symptoms are leading us in this direction. Never say never when it comes to infectious disease!"

"I'll need to get help with this," said the medical technologist. He made a call to a private laboratory.

"I can't do that testing without permission from the head of the Epidemiology Division at the National Institute for Communicable Diseases (NICD)," replied the supervisor at the private laboratory. "Dr. Blumberg must sign off on this."

"Do what you need to do, but don't waste any time," said the medical technologist. "If this is Ebola, we could have an outbreak on our hands. One person has died already."

The supervisor at the private laboratory made calls to the NICD.

"Ship the blood samples to us and Dr. Paweska will assay them in the **Biosafety Level–4 Laboratory**," replied Dr. Blumberg. Meanwhile, the paramedic was evacuated to the same hospital in Sandton a week after his symptoms began, where he died two days later.

Dr. Blumberg was on the phone with the physician who had treated the safari agent and paramedic who died. "Are there any other cases?" said Dr. Blumberg. "We need to find out who all had contact with the safari agent, who will be considered the **index case**."

That night, a nurse who had cared for the index patient was admitted to the hospital (case 3). Then two more cases were identified. One of them had cleaned the hospital room occupied by the safari agent. The cleaner (case 4) died 4 days after the paramedic (case 2). The fifth case was a nurse who had cared for the paramedic (case 2). All new cases were placed in an isolation ward. Four of the five individuals died. All of them suffered from a rapid deterioration, exhibiting respiratory distress, circulatory collapse, and neurological signs consistent with the index case.

The fifth case was still alive.

"The results are back from NCID, showing that the index case had a viral hemorrhagic fever," said one of the doctors as he observed the nurse who had cared for the paramedic, now suffering in isolation, lying on the hospital bed. "The virus is spread by fluid transmission such as contact with blood containing the virus. It is not Ebola as we had suspected but a virus related to the Lassa virus. Let's treat her with ribavirin. It has been used effectively to treat people infected with Lassa virus, and it's our only hope at this point."

"This appears to be more virulent than your typical hemorrhagic fever virus infection," said one of a team of healthcare workers standing over the last living case. The team of doctors stepped out of the isolation room and began to converse about the steps taken to identify the virus in this puzzling case.

The Special Pathogens Unit (SPU) in Sandringham, South Africa, worked with virologists at the Centers for Disease Control and Prevention (CDC) and Columbia University in New York. Dr. Paweska isolated the virus at the SPU, then sent them two postmortem liver biopsies from cases 2 and 3 and a serum sample from case 2. By isolating the virus, Paweska ruled out the need for other tests, including those for Ebola, Marburg, Lassa, Rift Valley fever, and Crimean Congo hemorrhagic fever.

Then the CDC and Columbia University researchers identified the virus to be a new strain of an *Old World* **arenavirus**.

Arenaviruses are usually associated with rodents. They are common in village homes and, therefore, human exposure to them can be frequent.

To make this identification, the researchers performed a procedure called **unbiased pyrosequencing** of the viral RNA extracts from the serum and liver biopsy tissues of the outbreak victims. Within 3 days of receiving samples, the researchers were able to identify a new virus related to Lassa virus, which is common in western Africa, and to lymphocytic choriomeningitis (LCM) virus, which is found all over the world. Prior to this outbreak, neither Lassa virus nor LCM virus had ever been found in southern Africa.

The doctors in Johannesburg adhered to practicing the barrier technique while they cared for the last living case suffering from the rare viral hemorrhagic fever in the isolation ward. This final patient (case 5) made a full recovery. Doctors didn't know if the ribavirin played a role in the patient's recovery or if the viral strain lost its virulence.

Update

The outbreak was contained to five people, which speaks volumes for the response teams and partnerships with NICD. It was fortuitous that the same doctor treated the first two victims. His suggestion to test the blood from the second victim for a viral hemorrhagic fever was very wise. Once the new virus was identified, it needed a name. Usually viral hemorrhagic fever viruses are named after a region or location. Thus, the new virus was named the Lujo virus, derived from the first two letters of the names of the cities in which it was found: Lusaka and Johannesburg.

Questions to Consider

1. Viral hemorrhagic fevers are caused by several distinct families of viruses. List these families and general characteristics of the viruses in these families (e.g., are they enveloped or not enveloped, composition of genetic material, host range, etc.)

2. Treatment with **convalescent serum** has been used successfully to treat some patients. Why wasn't that an option in this outbreak?

3. No vaccine prevents infection with the Lujo virus. What could have been done to prevent the infection in the safari travel agent?

4. Why was identifying the infectious agent causing this outbreak of great public health concern?

References

Briese, T., et al. 2009. Genetic Detection and Characterization of Lujo Virus, a New Hemorrhagic Fever—Associated Arenavirus from Southern Africa. *PLoS Pathogens* 4(5):e1000455.

Rodes, J. D., Salvato, M. S. 2006. Tales of Mice and Men: Natural History of Arenaviruses. *Revista Columbiana de Ciencias Pecuarias* 19(4):382–400.

World Health Organization. 2008. South African Doctors Move Quickly to Contain New Virus. *Bull WHO* 86(12): 912–913.

Internet Resources

National Institute for Communicable Diseases, A Branch of the National Health Laboratory Service, South Africa, http://www.nicd.ac.za/

Viral Hemorrhagic Fevers—CDC Special Pathogens Branch, http://www.cdc.gov/ncidod/dvrd/spb/mnpages/dispages/vhf.htm

Chapter 13

Deadly Wild Game Feasts

*Zoonotic diseases are caused by infectious agents that can be transmitted from animals to humans. Wildlife is a reservoir of infectious agents common to domestic animals and humans. Modes of transmission to humans include direct contact, bites (including insect bites), scratches, and consumption of contaminated meat or water. This is a study in which neurologists at the Neurobehavioral Institute located in Beaver Dam, Kentucky, turned to their medical records for answers after five unrelated patients came down with a **Creutzfeldt-Jakob disease (CJD)**-like illness. All five of the patients had a history of eating squirrel brains. Neurologists published a letter in the British medical journal* The Lancet *drawing attention to this pattern of illness.*

Neurologist Dr. Erick Weisman was offered a job in rural Kentucky in 1992 after finishing his internship and residency in Mobile, Alabama, and a fellowship at Boston University. He began practicing at a Kentucky clinic in 1993.

One day during his first year at the Kentucky clinic, Dr. Weisman read a patient's history before his next appointment. The patient was 54-year-old Harold Clementine. He was the mayor of his hometown in Kentucky and a salesman who had been laid off from his job for failing to complete all his paperwork (he had neglected to fill in the left side of sales forms). He had recently broadsided a car while driving because he claimed he couldn't see the car from the left. The first physician he saw after the collision diagnosed him with a stroke, which could have affected his peripheral vision on the left side. After extensive testing, it was later determined that Mr. Clementine actually suffered from a nondominant-hemispheric syndrome in which the left side of his brain had atrophied. This diagnosis was confirmed by magnetic resonance imaging (MRI).

"Mr. Clementine, can you draw a clock for me?" asked Dr. Weisman.

He watched the patient draw a circle with all of the numbers crammed into the right side of it. Mr. Clementine started to jerk like he had the hiccups. This formerly articulate, healthy man was beginning to deteriorate quickly from some kind of Cretuzfeld-Jakob disease–like neurological illness that was eating away at the left side of his brain. CJD is a rare, incurable neurodegenerative disorder caused by an infectious agent called a **prion**. While not a virus, prions are similar infectious agents in the sense that, like viruses, they are not "alive" and cannot replicate themselves outside of a host organism. Another similar disease, bovine spongiform encephalopathy (BSE) or "mad cow disease," became an epidemic in Great Britain, causing the death of more than 200,000 cattle during the 1980s to early 1990s. The BSE agent has been shown to cause a variant CJD in humans. Could consuming other animal products cause a different variant CJD?

Dr. Weisman knew that Mr. Clementine hunted squirrels and ate squirrel brains but didn't really think much about it. Hunting and eating squirrels is common in Kentucky. Dr. Weisman's wife, also a medical doctor, was born and raised in Kentucky. She grew up in the local culture.

In fact, in this region the brains of killed squirrels are considered a delicacy. When the famous cookbook, *The Joy of Cooking* by Irma S. Rombauer, was first published in 1936, it contained recipes of the Great Depression era, including recipes on how to prepare squirrel in dishes such as stews. One of the most popular ways to prepare squirrel brains is by scrambling them with eggs or in white gravy. The skull of the squirrel is cracked open and the brains are scooped out for cooking. Others prepare a stew called **burgoo** that contains squirrel brains, meat, and vegetables. People from all income levels eat squirrel brains in rural Kentucky, Texas, Alabama, Mississippi, Louisiana, and West Virginia (**Figure 7**)

Soon, more patients from the rural parts of Kentucky started showing up in Dr. Weisman's office with a CJD-like illness. As he contemplated these cases, Dr. Weisman

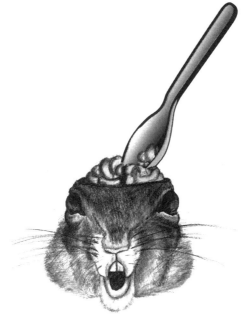

Figure 7. Prepared squirrel is a southern delicacy. (Drawn by Brian Ledwell.)

remembered a lecture he heard during his internship, given by Dr. Frank Bastian, a neuropathologist who suggested that *Spiroplasma* bacteria caused CJD. An intern had asked Dr. Bastian about the history of the patient, including his occupation and eating habits. Dr. Bastian had stated the patient ate squirrel brains.

After more reflection about Dr. Bastian's lecture and his current consultations with patients suffering from a CJD-like illness, Dr. Weisman called a colleague at the Department of Neurology at the University of Kentucky, Dr. Joseph Berger. Before his position in Kentucky, Dr. Berger was the Director of the CJD Foundation in Florida and had seen more than 20 cases of CJD.

"I've examined a cluster of patients suffering from a CJD-like illness," said Dr. Weisman. "Call me paranoid, but I am starting to wonder if there's a connection between their illness and eating squirrel brains."

Dr. Berger was intrigued. "Give me some more information on the background of the cases," he requested.

"None of these patients are related to each other and all of them lived in different rural communities," Dr. Weisman said. "Two of the patients are women and three are men. Their ages range from 56 to 78; making their average age of 68 at the onset of illness. All of the patients presented with symptoms within a three and a half year period."

"If you really think there is an association, we should do a retrospective study on the consultations and find out how many patients eat squirrel brains and what their initial diagnosis was," said Dr. Berger.

"Super idea!" said Dr. Weisman.

It didn't take that long for Dr. Weisman to review patient histories. There, he found a very interesting trend: 12 of 42 patients with Parkinson's disease who were seen at his clinic ate squirrel brains. The results intrigued Dr. Berger, who put together a joint letter for *The Lancet* that was published in 1997. He aimed to alert the medical community that there may be an association between squirrel brain consumption and a rare neurological illness. But there were still a lot of unanswered questions in this study.

"We still need to study the brains of squirrels for evidence of **prions** or *Spiroplasma*," said Dr. Berger, "though it's pretty well accepted in the medical and scientific community that prions cause CJD-like diseases rather than *Spiroplasma*."

"I've been encouraging hunters to send in squirrels for testing, including roadkills," said Dr. Weisman. "Unfortunately, the hunters haven't complied."

A year after the letter was published in *The Lancet*, the Kentucky state Department of Fish and Wildlife (DFW) killed 50 squirrels for study, but lab technicians refused to perform necropsies on the squirrels because they were afraid of potential exposure to a CJD-causing agent. Dr. Bastian agreed to necropsy the DFW squirrels and found no evidence of the sponge-like pathology in the squirrel brains indicative of CJD. Although disheartening, these results did not necessarily refute the possibility of "mad squirrel disease," because 50 specimens is not a large enough sample size in which to find a rare disease.

Meanwhile, the letter in *The Lancet* was not well received by the medical and scientific community. Drs. Berger and Weisman were highly criticized. Squirrel experts argued that the average life span of a squirrel is less than five years, making them a poor candidate to spread a disease with a long incubation period as spongiform encephalopathies tend to have. Furthermore, encephalopathies had never been reported in squirrels or in any wild animals that consume squirrels. Squirrels are also unlikely candidates for a prion disease, since other animals suffering from prion diseases likely got the disease from consuming prion-contaminated meat (for example, cows fed ground up spinal cords from carcasses of **downer cows**, who are exhibiting symptoms of BSE). Squirrels are primarily vegetarians, living on seeds, nuts, fruits, pinecones, leaves, twigs, and bark. (When faced with hunger and no other source of food, squirrels will eat bird eggs, insects, or pick at a carcass, and there are rare reports of male squirrels attempting to eat infant squirrels. These cases, however, are the exception to squirrel foraging behavior.)

Since the letter was published in *The Lancet* in 1997, only one more Kentucky squirrel-brain eater was diagnosed with a CJD-like illness in 1998. The authors didn't expect the volatile reaction from the medical and scientific community. *The New Yorker* magazine writer Burkhard Bilger interviewed the neurologists in 2000. His article on "Squirrel and Man" stated that Dr. Erick Weisman stands by his association of eating squirrel brains and a CJD-like illness. Dr. Berger was ambivalent. It is unlikely that *The Lancet* will publish a follow-up letter about squirrel brains. Interestingly, however, a new edition of *The Joy of Cooking* was published in 1997. The squirrel section was removed.

Update

Prion diseases have existed in animals for centuries. To date, there has been no convincing evidence that eating squirrel brains can cause a variant CJD. More studies are needed in search of a prion agent in the brains of squirrels to determine if there is an association between consumption of squirrel brains and a CJD-like illness.

Questions to Consider

1. List the symptoms of BSE.
2. List the symptoms of variant CJD.

3. List the symptoms of chronic wasting disease (CWD). Where is this disease most prevalent geographically?

4. List the hosts for BSE, variant CJD, and CWD.

5. What is Kuru? Where is this disease most prevalent geographically?

6. What is the treatment for CJD?

7. What experiments were done to prove that prions cause disease?

8. How does classic CJD differ from variant CJD?

References

Berger, J. R, Weisman, E., Weisman, B., 1997. Creutzfeldt-Jakob Disease and Eating Squirrel Brains. *Lancet* 350:642.

Bilger, B. Letter from Kentucky: Squirrel and Man; Is a Local Custom Worth Dying For? *The New Yorker*, July 17, 2000, pp. 59–67.

Hyman, H. L. 2002. Squirrel Brains: A Deadly Delicacy? *Postgrad Med* 112(1):12.

Kamin, M., Patten, B. M. 1984. Creutzfeldt-Jakob Disease: Possible Transmission to Humans by Consumption of Wild Animal Brains. *Amer J Med* 76:142–145.

Internet Resources

Home Page CDC Prion Diseases, http://www.cdc.gov /ncidod/dvrd/prions/

National Prion Disease Pathology Surveillance Center, http://www.cjdsurveillance.com/

Chapter 14

A Clinical Emergency

In 2007, patients may have been potentially exposed to bloodborne viruses at a local medical clinic in Nevada. Patients who had procedures at the clinic within a specific timeframe received letters recommending that they contact their primary care physicians or healthcare providers to get tested for a number of bloodborne viruses.

"I'm not feeling very well," said Joy Whittaker to her running partner one morning in November 2007. Joy loved living in Nevada. She could run year around without dealing with snow and harsh winters. She also enjoyed being in Las Vegas with all of its attractions from the MGM Grand to Caesars Palace.

"Wow, what happened to you?" replied her partner. "Your skin is a funny color; you look jaundiced. Are you eating too many carrots? I just read that too much beta carotene can make your skin change color."

"I don't eat that many carrots," responded Joy. "I just really feel wiped out and my belly hurts."

Joy Whittaker was very health conscious. She ate healthily and always had annual physical exams. She exercised. She was in her fifties and had started to wonder if the aging process was just starting to creep in. She made an appointment to see her primary care physician that day.

"Well, lab results show your liver enzymes are elevated," said her physician, Dr. Keech. Your alanine aminotransferase (ALT) levels are around 600 units per liter of serum. Normal ALT levels are in the range of 7 to 56 units per liter of serum. This is really odd for you, Joy. It is clear that you've got hepatitis, but my question is how? You don't have any risk factors for viral hepatitis unless it's hepatitis A? Have you been eating at the casinos a lot?" asked Dr. Keech.

"You know me," replied Joy. "I love going to the shows but I don't eat out much."

"Well, I'm going to have a full clinical work up done on you."

"Yuck! I don't need another endoscopy done, do I?" questioned Joy. "I just had that done a couple months ago at the clinic you recommended. That was no fun."

"Nope, no endoscopy. The lab will do several blood tests to see if you have an infection," said Dr. Keech. "I want to be thorough. In the meantime, don't do anything that may damage your liver. No alcohol, no Tylenol, and get plenty of rest. Eat a high protein/carbohydrate diet to help repair any damaged liver cells and protect your liver."

Joy went home and followed the doctor's orders. A few days later, she got a call from Dr. Keech to come in for another appointment.

Joy's jaw dropped when she heard the diagnosis. "I'm infected with hepatitis C virus?" she asked. "Where could I have picked that up?"

"It's most commonly transmitted through contaminated needles to inject drugs," responded Dr. Keech. "I know you don't use drugs."

"There's got to be a logical explanation for this," said Joy.

"Exactly," affirmed Dr. Keech. "Let's focus on treatment right now, and we'll get to the bottom of the cause once you're on the mend. I am going to treat you with a combination of drugs, **ribavirin** and **pegylated interferon**," said Dr. Keech. "There are at least six different genotypes of hepatitis C virus. You've got the most common type in the United States—genotype 1. Most people infected with genotype 1 have undetectable levels of hepatitis C virus RNA in their blood after 24 weeks of treatment."

"I know this is difficult," continued Dr. Keech. "In the meantime, I will ask the lab to repeat tests—ruling out hepatitis A and B and other causes of hepatitis. The hepatitis C serology testing and RNA assays will also be repeated. As you said, there has to be a logical explanation for how this infection occurred, assuming this is the correct diagnosis," said Dr. Keech.

Two days later, another jaundiced patient showed up in Dr. Keech's office, another woman in her late fifties. Dr. Keech required her to be hospitalized. He scrutinized her medical records very carefully. Her test results were the same—she had a hepatitis C virus (genotype 1) infection. Like Joy Whittaker, this patient had also had an endoscopy procedure a couple of months ago at the same clinic. Dr. Keech was immediately on the phone to the Endoscopy Center.

"Call it coincidence but two of my patients have contracted hepatitis C virus," said Dr. Keech. "Both of them are females in their fifties and both had an endoscopy procedure at your clinic two months ago on the exact same day," said Dr. Keech. "Neither of these patients have any risk factors for contracting hepatitis C virus."

"Thank you for your concern," said Dr. Desai. "I will check into these specific cases and get back to you."

A month later, a jaundiced 72-year-old male showed up in Dr. Ted Langland's Las Vegas walk-in clinic. The man required hospitalization. Dr. Langland ordered a battery of tests, including ALT levels and serology tests to detect the presence of hepatitis virus antigens, ALT, and bilirubin levels. He had been this man's primary care physician for 30 years. The man had no risk factors for contracting hepatitis B or C viruses. A few days later, the man was diagnosed with a hepatitis C virus infection.

Dr. Langland and Dr. Keech were colleagues and friends. They golfed together every Wednesday afternoon. They rarely talked shop, but Dr. Keech was very bothered by the fact that two of his patients were recently infected by hepatitis C virus and neither one had any risk factors. The subject came up with Dr. Langland.

"Did your patient have an endoscopy procedure at Dr. Desai's clinic too?" asked Dr. Keech. When Dr. Langland responded in the affirmative, Dr. Keech exclaimed, "I don't believe this!"

The two of them were astonished and deeply concerned. After their golf game, both of them were on the phones to the Endoscopy Center.

"We are aware of the situation Dr. Keech, our records are being surveyed and cross-matched with a local laboratory's records of procedure records done at this clinic," said Dr. Desai. "It appears that six people with no obvious hepatitis C infection risk factors have contracted the hepatitis C virus. All six of these had an endoscopy procedure done on the same day or same week at this clinic."

During the days that followed, health officials observed procedures at Dr. Desai's endoscopy clinic. The clinic typically performed 50 to 60 endoscopy or colonoscopy procedures a day. The clinic was open 5 days a week. Nevada state health officials observed staff reusing syringes on individuals and saw the use of medication vials intended for single use being used on multiple persons. Patients were exposed to the blood of others when the staff reused needles and vials of anesthesia (**Figure 8**).

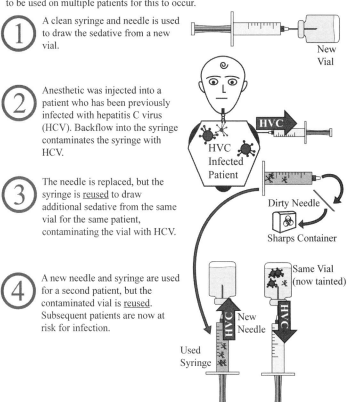

Reuse of syringes combined with the use of single-dose vials for multiple patients undergoing anesthesia can transmit infectious diseases. The syringe does not have to be used on multiple patients for this to occur.

1. A clean syringe and needle is used to draw the sedative from a new vial.

New Vial

2. Anesthetic was injected into a patient who has been previously infected with hepatitis C virus (HCV). Backflow into the syringe contaminates the syringe with HCV.

HVC

HVC Infected Patient

3. The needle is replaced, but the syringe is <u>reused</u> to draw additional sedative from the same vial for the same patient, contaminating the vial with HCV.

Dirty Needle

Sharps Container

4. A new needle and syringe are used for a second patient, but the contaminated vial is <u>reused</u>. Subsequent patients are now at risk for infection.

Same Vial (now tainted)

New Needle

Used Syringe

Figure 8. Unsafe injection practices and disease transmission. (Adapted from Labus, B. et al., 2008. *Acute Hepatitis C Virus Infections Attributed to Unsafe Injection Practices at an Endoscopy Clinic—Nevada, 2007.* Drawn by Brian Ledwell.)

After the exposure of these unsafe practices, the clinic issued a statement saying it had cleaned up its act and was cooperating with the health district's investigation.

Update

About 63,000 persons were identified in the above incident as being at potential risk for acquisition of a bloodborne pathogen. More than 50,000 patients were advised to be tested, and at least 115 patients were infected with hepatitis B virus, hepatitis C virus, or HIV. A civil suit was filed in the Eighth Judicial District Court of Nevada with allegations of gross negligence. Punitive damages were being sought. The investigation is ongoing. More than 35 outbreaks of viral hepatitis have occurred in the United States in the past 10 years as a result of unsafe injection practices, including syringe reuse among patients, contamination of medication vials or intravenous bags, and inappropriate sharing of blood sugar testing equipment.

Questions to Consider

1. List safe injection and medication procedures used in the practices of healthcare.

2. Compare and contrast hepatitis A, B, and C viruses. How is each type of viral infection treated?

3. What is **cirrhosis** of the liver?

4. Can all three of the above hepatitis viruses cause a chronic infection? Define a **chronic infection**.

5. What is an **acute** hepatitis virus infection?

6. What is a **chronic** hepatitis virus infection?

7. Unsafe injections are suspected to occur routinely in developing countries. Why? Do a literature search to support your answer.

8. Compare and contrast the modes of transmission for hepatitis A, B and C viruses.

9. What are risk factors for infection by hepatitis A, B, or C viruses?

10. Research and list geographic locations where the six different hepatitis C virus genotypes are most common.

11. Why are there fewer cases of hepatitis A and B infection today than 30 years ago?

12. What other viruses besides hepatitis A, B, and C can cause hepatitis?

13. List some nonviral causes of hepatitis.

14. Create a list of bloodborne pathogens.

References

Black, L. M. 2011. Tragedy into Policy: A Quantitative Study of Nurses' Attitudes Toward Patient Advocacy Activities. *Am J Nurs* 111(6):26–35; quiz 36–37.

Feinmann, J. 2010. Doctors Call for Ban on Multidose Vials after Hepatitis C Outbreak in US. *BMJ* 341:c4057. doi:10.1136/bmj.c4057.

Fischer, G. E., et al. 2010. Hepatitis C Virus Infections from Unsafe Injection at an Endoscopy Clinic in Las Vegas, Nevada 2007–2008. *Clin Infect Dis* 51(3):267–273.

Labus, B., et al. 2008. Acute Hepatitis C Virus Infections Attributed to Unsafe Injection Practices at an Endoscopy Clinic—Nevada, 2007. *MMWR* 57(19):513–517.

Pandit, N. B., Choudhary, S. K. 2008. Unsafe Injection Practices in Gujarat, India. *Singapore Medical Journal* 49(11):936-939.

Reid, S. 2009. Increase in Clinical Prevalence of AIDS Implies Increase in Unsafe Medical Injections. *Internat J STD & AIDS* 20:295–299.

Simonsen, L., et al. 1999. Unsafe Injections in the Developing World and Transmission of Bloodborne Pathogens: A Review. *Bulletin of the World Health Organization* 77(10): 789–800.

Thompson, N. D., et al. 2009. Nonhospital Health Care—Associated Hepatitis B and C Virus Transmission, 1998–2008. *Ann Intern Med* 150:33–39.

Internet Resources

CDC Viral Hepatitis, http://www.cdc.gov/hepatitis/

APIC Position Paper: Safe Injection, Infusion and Medication Vial Practices in Healthcare, http://www.apic .org/Resource_/TinyMceFileManager/Position_Statements /AJIC_Safe_Injection0310.pdf

Glossary

Acute infection Infections that have symptoms of sudden onset or last a short time.

AIDS (acquired immunodeficiency syndrome) A serious viral disease caused by the human immunodeficiency virus (HIV) in which the T (CD4) lymphocytes are destroyed and opportunistic illnesses occur in the patient.

Alanine aminotransferase (ALT) An enzyme in the body that, when at elevated levels, indicates liver inflammation that can be the result of viral hepatitis.

Amantidine A synthetic antiviral drug.

Antibody A highly specific protein produced by the body in response to a foreign substance, such as a bacterium or virus, and capable of binding to that substance.

Antigen A chemical substance that stimulates the production of antibodies by the body's immune system.

Antiviral drug An agent used to treat viral infections that can inhibit or destroy viral replication.

Asian tiger mosquito Day-biting mosquito that is a carrier of a number of viruses that cause encephalitis.

AZT (Zidovudine) The first antiviral drug approved for HIV-1 treatment. It is a thymine analog that inhibits reverse transcription of viral genomes.

Barrier technique A method of containing infection in which an impenetrable object or surface is placed between the infected and the uninfected or body (e.g., gloves, gowns, and masks worn to prevent the spread of infection).

Bilirubin A yellow-pigmented byproduct of old red blood cells that causes jaundice.

Biopsy The removal of a sample of tissue for purposes of diagnosis.

Biosafety Level 4 Indicates that the strictest containment and safety measures possible are in place for facilities that work with and study dangerous microorganisms and viruses that pose a high risk of life-threatening disease and for which there is no cure, treatment, or vaccine available (e.g., the Ebola virus).

BSE (bovine spongiform encephalopathy) Commonly known as "mad cow disease," is a fatal neurological disease that causes a spongy degeneration of the brain.

Burgoo Thick spicy stew that contains whatever meat (e.g., squirrel brains) and vegetables are available; often served as a delicacy in the southern United States.

CDC (Centers for Disease Control and Prevention) U.S. agency located in Atlanta, Georgia; charged with tracking and investigating infectious diseases and public health trends.

Chronic infection An infection lasting more than six months.

Cirrhosis The formation of fibrous tissues, nodules, and scarring on the liver that interfere with its cell functions and blood circulation.

CJD (Creutzfeldt-Jakob disease) A rare, fatal, neurological disease hypothesized to be caused by a prion that alters the structure of a normal protein, resulting in the destruction of brain tissue.

Climate change Any long-term change in the weather patterns over periods of time that rang from decades to millions of years.

Coma A state of prolonged and deep unconsciousness as a result of disease, injury, or medical procedure.

Convalescence The recovery period after an illness.

Convalescent serum Antibody-rich serum obtained from a convalescing patient.

CSF (cerebrospinal fluid) The colorless fluid in and around the brain and spinal cord that absorbs shocks and maintains uniform pressure.

CT scan A radiological scan in which cross-sectional images within a part of the body are formed using computerized techniques and are shown on a computer screen.

Cyanosis Bluish discoloration of the skin and mucous membranes caused by a lack of oxygen in the blood. It can be associated with certain viral diseases (e.g., blue ear pig disease, blue lips of Ebola patients).

DEET The active chemical ingredient in many insect repellents that are applied to the skin.

Dengue fever Serious illness caused by a dengue virus carried by *Aedes aegypti* mosquitoes and most often found in hot climates.

Dengue hemorrhagic fever Serious illness caused by the dengue virus resulting in abdominal pain, hemorrhage (bleeding), and circulatory collapse (shock).

Diagnosis The identification of an illness or disorder in a patient through an interview, physical examination, medical tests, and other procedures.

Differential diagnosis A systematic method used by clinicians to identify disease causing a patient's symptoms.

Downer cows Live cow that cannot walk due to disease or injury. Some Downer cows may be suffering from BSE.

Electrolytes Ions [sodium (Na^{2+}), potassium (K^+), calcium (Ca^{2+}), magnesium (Mg^{2+}), chloride (Cl^-), phosphate (PO_4^{3-}), and bicarbonate (HCO^{3-})] found in cells or the blood.

ELISA (enzyme-linked immunosorbent assay) An assay that relies on an enzymatic conversion reaction to detect the presence of antigen or antibody in a sample.

Encephalitis Inflammation of the tissue of the brain or an infection of the brain.

Encephalopathy General term for a disease of the brain.

Endoscopy Use of an endoscope to examine a bodily orifice, canal, or organ.

Epidemic An unusually high number of cases in excess of normal expectation of a similar illness in a population, community, or a region.

Etiological agent The disease causing agent.

GARDASIL vaccine Vaccine designed to prevent infection by certain high-risk types of human papillomaviruses (HPV) such as HPV 16, 18, 6, and 11 that cause a large percentage of cervical, anal, vaginal, vulvar, and penile cancers.

Gastroenteritis An inflammation of the stomach and the intestines, usually as a result of a bacterial or viral infection.

Genbank database An open-access sequence database that contains nucleotide and protein translations for more than 100,000 distinct organisms.

Genotype A group of organisms or viruses that share a specifice genetic constitution.

Global warming An increase in the average temperature of the Earth's atmosphere resulting from human activities such as pollution production.

Hematocrit A blood test that measures the percentage of red blood cells found in whole blood.

Hepatitis Inflammation of the liver caused by a number of viruses, alcohol, or prescription drugs.

HIV (human immunodeficiency virus) Either of two strains of a retrovirus, HIV-1 or HIV-2, that destroys specific groups of immune cells, the loss of which causes AIDS.

Host An organism on or in which a microorganism lives and grows or in which a virus replicates.

Hypersalivating Drooling or slobbering generally caused by excess production of saliva (symptom of rabies infection).

IgM antibodies A class of antibodies that is the first immunoglobin to respond to a viral or other microbial antigen.

Immunofluorescent staining A method of detecting viral antigens within a cell or on its surface. The virus-specific antibodies containing a fluorescent tag bound to the FC region of the antibody are allowed to react with the specimen containing the virus and any unbound antibody is washed away and the specimen is observed under a fluorescent microscope.

Index case "Patient zero" or the first documented case of a disease in an epidemiological study.

Inflammation The swelling, redness, heat, and pain produced in an area of the body as a reaction to injury or infection.

Intravenous fluids therapy Giving liquid substances directly into a vein.

Jaundice Yellowing of the skin and eyes caused by too much bilirubin in the blood.

Ketamine General anesthetic (can be used to induce coma as in the Jeanna Giese rabies case).

Malaria An infectious disease caused by the parasite *Plasmodium*, which is transmitted by the bite of infected female mosquitoes and is characterized by recurring chills and fever; also called blackwater fever.

Microarray analysis Method that allows scientists to detect thousands of genes simultaneously in a small sample and to analyze the expression of those genes.

Milwaukee protocol Experimental course of treatment used on Jeanna Giese (had a rabies infection and did not receive post-exposure vaccination).

Modes of transmission Defines how an infectious disease is spread or passed on; can be direct or indirect.

MRI (magnetic resonance imaging) Uses nuclear magnetic resonance of protons to produce proton density images of the soft tissues of the body.

Médecins Sans Frontières (MSF) or Doctors Without Borders A humanitarian organization that brings quality medical care to people whose survival is threatened by violence, neglect, or a catastrophe primarily due to armed conflict, epidemics, malnutrition, exclusion from health care, or natural disasters.

Necrosis The death of an area of living tissue or cells.

Necropsy The examination or autopsy on a dead animal.

Nystagmus Involuntary movements of the eye (symptom of rabies infection in humans).

Old World arenaviruses Arenaviruses present in the eastern hemisphere of the world that was known to Europeans before the discovery of the Americas.

Outbreak A small, localized epidemic.

Pegylated interferon A conjugate of recombinant interferon alpha and polyethylene glycol (PEG) used as an antiviral (e.g., hepatitis C treatment) or anticancer drug.

Pre-mastication A practice in which mothers or caregivers chew food before feeding it to an infant.

ProMED mail Global electronic reporting system for out-breaks of emerging infectious diseases and toxins.

PCR (polymerase chain reaction) A technique used to rep-licate a fragment of DNA and produce very large quantities of that sequence.

Reservoir Where the etiological agent lives, grows, and multiplies.

Roundup® Chemical weed killer.

Prion An infectious, self-replicating protein containing no genetic material that is responsible for a number of neurode-generative brain diseases in both humans and animals.

Re-emerging viral diseases Diseases caused by viruses that are making a "comeback" or are reappearing and causing increased incidence or geographic range of infections in exposed human populations.

Ribavirin Antiviral drug used to treat severe respiratory syncytial virus (RSV) infections, hepatitis C infections in conjunction with pegylated interferon, and other viral infec-tions (e.g., Milwaukee protocol).

RT-PCR (Reverse Transcriptase-Polymerase Chain Reaction) PCR protocol in which cDNA is made from RNA by reverse transcription; the cDNA is then multiplied by standard PCR protocols. Used in tests to measure HIV or hepatitis C viral load and other diagnostic testing of viruses with RNA genomes.

Septic shock A serious medical condition that occurs when an overwhelming infection leads to low blood pressure and low blood flow; leading to decreased oxygen delivery, it can cause multiple organ failure and death.

Seroconverted The development of detectable specific anti-bodies to viruses or microbes in the blood serum as a result of an infection.

Sign An indication of the presence of a disease or disorder, especially one observed by a doctor but not apparent to the patient (e.g., low-grade fever, high blood pressure).

Symptom An indication of some diseases or disorder, espe-cially one experienced by the patient (e.g., pain, headache, or itching).

Symptomatic treatment Treating the symptoms of an ill patient.

Syndrome A group of signs and symptoms that, taken together, characterize a specific disease or disorder.

Transfusion The injection of blood or blood plasma into a vein or artery.

Transmission electron microscopy (TEM) Microscopic technique in which a beam of electrons (instead of light, which is used in light microscopy) is transmitted through an ultra thin specimen. An image is formed from the interaction of the electrons and the specimen. It is used to visualize viruses because it has a resolving power in the nanometer range—the size of viruses.

Transplant Operation in which an organ from a donor is moved to a recipient (as in "she had a kidney transplant")

Traveler's diarrhea A common illness affecting travelers who have three or more unformed stools in 24 hours, commonly accompanied by abdominal cramps, nausea, and bloating.

Unbiased high-throughput sequencing A fast method of determining the order of bases in DNA.

Unbiased pyrosequencing A technique used determine the order of bases in DNA using chemiluminescent enzymatic reactions.

Vaccine A preparation containing weakened or dead microorganisms or viruses, treated toxins, or parts of microorganisms or viruses to stimulate immune resistance.

Virus neutralization assay A type of antigen-antibody reaction in which the activity of a virus is inactivated.

Wart A small, usually benign, skin growth commonly due to a virus.

Index